NATURE AND BUREAUCRACY

This book questions how bureaucracies conceive of, and consequently interact with, nature, and suggests that our managed public landscapes are neither entirely managed nor entirely wild, and offers several warnings about bureaucracies and bureaucratic mentality.

One prominent challenge facing scientists, policymakers, environmental activists, and environmentally concerned citizens, is to recognize that human influence in the natural world is pervasive and has a long history. How we act, or choose not to act, today will continue to determine the future of the natural world. Western-style management of nature, mediated by economic rationality and state bureaucracies, may not be the best strategy to maintain environmental integrity. The question is, what kinds of human influence, conceived of in the widest possible sense, will produce ideal environments for future generations? The related question is, who gets to choose? The author approaches the problem of analyzing the mutual influence of human and natural systems from two perspectives: as an objective scholar investigating bureaucracies and natural systems from the outside, and over the last decade as an inside practitioner working in various roles in federal land management agencies developing policies and regulations involved in the control of natural systems.

This book will be of great interest to students and scholars of natural resource management, policy and politics, and professionals working in environmental management roles as well as policymakers involved in public policy and administration.

David Jenkins has 12 years of experience working in U.S. land management agencies. Prior to that he taught at the Massachusetts Institute of Technology and Bates College and conducted research at the Bureau of Applied Research in Anthropology at the University of Arizona. His research publications span a range of topics, including myth, social organization, kinship, exchange networks, museums, ethnographic photography, environmental values, endangered species, resource exploitation, subsistence fisheries, autobiography, and the use of mathematical models in anthropology.

ROUTLEDGE EXPLORATIONS IN ENVIRONMENTAL STUDIES

For more information about this series, please visit: www.routledge.com/ Routledge-Explorations-in-Environmental-Studies/book-series/REES

NATURE AND BUREAUCRACY

The Wildness of Managed Landscapes

David Jenkins

LONDON AND NEW YORK

from Routledge

Cover image: Getty

First published 2023
by Routledge
4 Park Square, Milton Park, Abingdon, Oxon OX14 4RN

and by Routledge
605 Third Avenue, New York, NY 10158

Routledge is an imprint of the Taylor & Francis Group, an informa business

British Library Cataloguing-in-Publication Data
A catalogue record for this book is available from the British Library

Library of Congress Cataloging-in-Publication Data
Names: Jenkins, David (Anthropologist), author.
Title: Nature and bureaucracy : the wildness of managed
landscapes / David Jenkins.
Description: New York : Routledge, 2023. |
Includes bibliographical references and index.
Identifiers: LCCN 2022009201 (print) | LCCN 2022009202 (ebook)
Subjects: LCSH: Natural resources–Management. |
Public administration. | Bureaucracy.
Classification: LCC S944.5.D42 J46 2023 (print) |
LCC S944.5.D42 (ebook) | DDC 639.9068–dc23/eng/20220330
LC record available at https://lccn.loc.gov/2022009201
LC ebook record available at https://lccn.loc.gov/2022009202

ISBN: 978-1-032-28567-2 (hbk)
ISBN: 978-1-032-28562-7 (pbk)
ISBN: 978-1-003-29744-4 (ebk)

DOI: 10.4324/9781003297444

Typeset in Bembo
by Newgen Publishing UK

Drought

What will we do when the well's dry?
When the water sinks into
sand, will we ask the sky
with swollen lips split in two?

When the water sinks into
deeper aquifers we'll whisper, dread,
with swollen lips split in two
red cracked riverbeds.

"Deeper aquifers?" we'll whisper. Dread.
Even moss can find life in
red cracked riverbeds.
Our roots are short and will not lengthen.

Even moss can find life in
red cracked riverbeds.
Our roots are short and will not lengthen.
For strength we have only grief.

Between the bricks we leave
sand. Will we ask the sky
for strength? We have only grief.
What will we do when the wells dry?
                          —*Ravn M. Jenkins*

CONTENTS

FIGURES

ACKNOWLEDGEMENTS

Chapter 2 first appeared in *Environment and History*, Volume 15, Number 4 (November 2009). Chapter 3 first appeared in *Comparative Studies in Society and History*, Volume 45, Issue 4 (October 2003). Chapter 4 first appeared in *Marine Policy*, Volume 61 (November 2016). My thanks for the permission to reprint these studies, which have been modified and updated for inclusion here.

Thanks to Ravn M. Jenkins for her pantoum, "Drought," a difficult poetic form that captures the recursion of nature's modification by humans, and the consequences.

I must also thank family, friends, teachers, and the many bureaucrats I've worked with over the last decade in the U.S. Fish and Wildlife Service, the U.S. Forest Service, and the Bureau of Land Management. All have challenged—and improved—my understanding of public resources. All have tolerated, if sometimes barely, my insistent and incessant questioning of the assumptions underlying management of our shared landscapes.

Historian Richard White may see his influence in these pages. My lifelong debates with evolutionary biologist Dave Carrier about humans and wildlife also appear in these pages, though in disguised form. Anthropologist Bojka Milicic prodded me to keep working, as she puzzled over my shift in focus. Relational biologist Robert Bugg's enthusiasm for ideas and their practical application continues to sharpen my own sense of possibilities. Francisco Valenzuela provided timely advice on how to negotiate space to make decisions in situations of bureaucratic folly. Essayist, poet, and anthropologist Robin Fox insisted I keep my sense of humor and reminded me never to confuse seriousness with solemnity. His reminder called to mind philosopher Susanne Langer's remark that scholars are engaged with showing their mudpies to one another, after all. What could be more playful, and serious, than that? As always, thanks to Mona Letourneau for her careful editing, which brought not only clarity, but also beauty and substance to my muddy prose.

INTRODUCTION

The Wild Garden

One prominent challenge facing scientists, policymakers, environmental activists, and environmentally concerned citizens is to recognize that human influence in the natural world is pervasive and has a long history and to act accordingly—or to choose not to act. Western-style management of nature, mediated by economic rationality and state bureaucracies, may not be the best strategy to maintain environmental integrity. The question is, what kinds of human influence, conceived of in the widest possible sense, will produce ideal environments for future generations? The next question is this: who gets to decide on a course of action? *Nature and Bureaucracy* explores these questions and places modern bureaucratic culture and function smack in the middle of them. We may be in the age of total bureaucratization, but bureaucrats do not entirely control our cultural meanings, nor our messages, nor, fortunately, the mix of natural and cultural systems I call our wild gardens.

How should we describe and analyze the synergies of human and natural systems? Governmental bureaucracies are particularly important in defining that mutual influence with a range of consequences, some intended, some not. I've approached the problem of analyzing the mutual influence of human and natural systems from two perspectives, first as an objective scholar investigating bureaucracies and natural systems from the outside, and over the last decade, secondly, as an inside practitioner working in various roles in federal land management agencies developing policies and regulations involved in the control of natural systems. In both instances, I've been guided by anthropologists, historians, sociologists, political scientists, and ecologists to demonstrate how states and their bureaucracies are intertwined with natural systems.

The key questions pursued here are, how do bureaucracies conceive of and consequently interact with forests? With water? With fish? With wildlife? With humans? Pursuing these questions calls for a description of "traditional bureaucratic

DOI: 10.4324/9781003297444-1

knowledge." First, we understand traditional bureaucratic knowledge not through what bureaucrats say, but through what they do. Bureaucratic knowledge becomes visible in action—for example in the building of a system of canals, or cutting a forest in a patchwork pattern, or using helicopters to round up thousands of wild horses on public lands, or drilling for oil in Alaska wilderness. Herbert Kaufman, in his classic study of the U.S. Forest Service, made the same point: "Concrete actions … add up to the sum and substance of Forest Service policy; everything else is only intention."[1]

Action is only part of the story, however, and perhaps the least interesting part, even though it receives the most critical attention. This leads to my second claim. Traditional bureaucratic knowledge is always uncertain because it resides *between* knowledge systems, in particular between economic, scientific, political, and cultural knowledge systems. It is not a monolith, and can never be a monolith, although bureaucracies have come to dominate much of our social life. Kaufman's dismissive "only intention" misses much, since "intention" flags social and cognitive phenomena that motivate behavior and, consequently, are of interest to land use planners, economists, nongovernmental organizations, and politicians, among many others. Motivations of behavior include symbolic systems, power relations, expressions of morality, concentrations of significance, and the general, ongoing contestation of meaning. Each chapter presents the issue of cultural intentionality and the contestation of meaning in the context of public resources management, moving beyond a singular focus on concrete action.

Because I see bureaucratic knowledge as residing between knowledge systems, I find sociologist Max Weber's pioneering work less salient than more recent scholarship. As social forms, bureaucracies are considerably more mixed, and mixed up, than Weber imagined. Still, over a century ago, Weber identified common patterns as large-scale government and business bureaucracies emerged and solidified their hold on us. I'll mention five common patterns.

First, characterized by a "rigid division of labor" and "necessary chains of command," bureaucracies are staffed by those with "certified qualifications." The result is a self-perpetuating hierarchically ranked organization resistant to change as certified experts become embedded in chains of command. Second, "bureaucratic administration … produces an optimal efficiency for precision, speed, clarity, command of case knowledge, continuity, confidentiality, uniformity, and tight subordination." Optimal efficiency becomes a transcendent cultural value which is rarely questioned. Instead, it is pursued with great administrative zeal. Third, "The predictability of rules is, for a modern bureaucracy, actually of primary importance." Without predictability, and without rules, bureaucratic administration would fail to reach its transcendent value of optimal efficiency, further frustrating those at the top of the chain of command. Fourth, "secrecy is used to sustain the power interests of the bureaucracy." Indeed, "nothing is more fanatically protected by bureaucracy than the concept that secrecy is necessary…" Fanatically protected secrecy may be of bureaucratic concern, as those higher in the chain of command solidify their power by excluding others from decision-making. But in public lands management, indeed in many functions of government, secrecy runs counter to public benefit and immediately sets up a conflict, which further implicates diverse knowledge systems. Fifth, "unfailingly, a fully developed bureaucracy is always in a position of great power." This is both true and overstates the case, as the recent history of certain government bureaucracies in the United States demonstrates, the Environmental Protection Agency and the Department of Agriculture among them, which lost influence as they were partially dismantled in response to political direction.[2]

In the following pages, I show that these Weberian characteristics of bureaucracies are often undercut by the people who inhabit them, but not entirely, or at least not entirely successfully. For example, I address the presumption of bureaucratic efficiency in the chapter on timber harvests and forest restoration. "Efficiency" became both a measure of effort and a moral value, which in combination served to preclude scrutiny of either of them. An examination of the actual knowledge systems at play, however, shows that things are less efficient and much messier than the bureaucracy itself intends. Similarly with the presumption of secrecy. Land management bureaucracies in the United States are more porous than their leaders would like, but porous they are and for good reason. Decisions about public lands should always be entirely public, even if those in power endeavor to keep their decisions hidden. Yet in any conflict over openness, those in charge of a bureaucracy can get rid of subordinates who value transparency over secrecy. Many scientists have been summarily booted from their positions in land management agencies for continuing to publicly describe scientific findings in contradiction to political preference. Less drastically, those in charge of a bureaucracy can simply ignore scientific or other knowledge systems. This has happened, for example, during the political apportionment of Colorado River water between states in the western United States. Bureaucrats used measurements of historic river flows known to be flawed but used them anyway. In such cases, specialists with "certified qualifications" may continue to publish their findings, but those in control of this or that bureaucracy may refuse to accept them.

According to political scientist James C. Scott, one additional characteristic of modern states and their constituent bureaucracies is the need to radically simplify an understanding of natural and social worlds. This is necessary to make those worlds legible to decision makers. Unanticipated problems arise, however, when significant aspects of natural and social worlds lie outside the narrowed field of vision of the state. A crisis may be the only way to broaden that vision. Climate change is one such crisis, while catastrophic wildfire is another.

Scott also notes:

> The modern state, through its officials, attempts with varying success to create a terrain and a population with precisely those standardized characteristics that will be easiest to monitor, count, assess, and manage. The utopian, immanent, and continually frustrated goal of the modern state is to reduce the chaotic, disorderly, constantly changing social reality beneath it to something more closely resembling the administrative grid of its observations.[3]

Such administrative grids have come to be expected and are as common as disorderly social and natural worlds. And yet the possibility of transformation remains a challenge to both bureaucracies and individuals, as the disorderly swirl becomes captured by the observational grid. The wild garden runs the risk of becoming a thoroughly regimented industrial farm of corn or soybeans. Anthropologist Roy Rappaport often observed that humans live in a world of causal law, but in terms they themselves construct. We must therefore pay attention to cause and effect and meaning, to physical law and cultural expression, and seek to understand the complex dynamic of their interaction. Bureaucracies are one such site of interaction, whose dynamics I seek to understand if not transform.

To analyze the functioning of a governmental bureaucracy, one must look to its effects on the natural world. At the same time, analyses of the natural world inevitably lead to a consideration of bureaucratic culture and practice. As noted in the journal *Science*, "most aspects of the structure and functioning of the earth's ecosystems cannot be understood without accounting for the strong, often dominant influence of humanity."[4] For good and ill, that dominant influence has been bureaucratic.

I first entered this arena of scholarship more than 25 years ago. I studied the relationships between nature, government, and bureaucratic process as part of a large international research team. My focus was on oilfield waste from the Gulf of Mexico and on the promise of sustainable building practices in southern Arizona. The result was a very good, very fat collaborative book on the nexus of local cultural values and large-scale environmental policies in the United States, India, Japan, and China, published in 2006 and entitled *Forging Environmentalism: Justice, Livelihood, and Contested Environments*.[5] After that, I studied the regulatory process of listing Atlantic salmon as an endangered species, and later turned to the long history of aquifer depletion and the troubled use of the Colorado River in the American Southwest.[6] These studies are reflected in *Nature and Bureaucracy*.

Since then, I moved out of full-time teaching and research and into a job working for the U.S. Fish and Wildlife Service in Alaska. At that point, I had the opportunity to analyze the relationships between nature, government, and bureaucracy from the inside. One result was a study of the vast Yukon River fishery and the demise of Chinook salmon.[7] Another was an analysis of how anthropology became co-opted by state and federal bureaucracies. My position as a bureaucrat took various roles, anthropologist, fisheries supervisor, policy coordinator, acting deputy regional director of migratory birds and state programs, among others. These appointments allowed me to participate in the formation of regulations governing human uses of the natural world. I thought it important to reflect on and report what I had learned as an insider to that process, and I include both studies here.

After nearly five years in Alaska, I shifted jobs and agencies to work for the U.S. Forest Service. My new set of responsibilities included directing the recreation, wilderness, archaeology, and volunteer programs across 14 national forests and one national tallgrass prairie in the eastern United States. I went from working in a huge remote Alaskan landscape dominated by public lands and with fewer than 800,000 people state-wide, to working in a landscape with relatively tiny public lands but with over 40 percent of the U.S. population.

I became interested in trees, in part as a hazard of working for the Forest Service, and in part because most of the national forests in the eastern part of the United States were regrown after earlier and devastating practices of clearcutting. The regrowth of eastern forests was a signal success, initiated under the 1911 Weeks Act, which was enacted to protect threatened watersheds. Well over a century ago it was apparent that clearcutting forests resulted in massive erosion and uncontrollable floods. The federal government bought up land to protect eastern watersheds, which became national forests.[8] Recently, under the Trump administration, our marching orders were to increase timber production by 15 percent in the national forest system. As it turns out, the eastern region is one of the two or three most productive timber regions for the Forest Service, the others being the southern and northwestern regions. Once again, I thought it important to reflect on and report what I learned as an insider to that process.

As an outside scholar and inside practitioner, I've sought to maintain a level of objectivity. Objectivity is not entirely possible in ethnographic research, as a large anthropological literature demonstrates, but I've never been overly troubled by that demonstration.[9] I have learned nonetheless that within a bureaucracy to remain thoughtful requires one to remain marginal. I find as soon as I start using bureaucratic lingo unselfconsciously, then it is time to readjust my ethnographic awareness. Participant observation in anthropological research is one thing; becoming socialized is quite another. Working within a large bureaucracy carries its own challenges for anyone trained in the human sciences. The way I best readjust my awareness is through writing.

I have also sought to ignore objectivity, in the sense that some of my responses to working within a bureaucracy are personal. I could not simply pass over this aspect of my experience. I've included a few of those responses in the hopes of providing a different sort of insight into the workings of land management bureaucracies. I've never been interested in making a work of scholarship about myself, directly or indirectly, but have allowed a bit more of my personal responses to emerge in this context.

In some ways, a more personal response is needed to counter what Weber called "a fully developed bureaucracy" which "embodies very specifically the principle of *sine ira ac studio* (i.e., without anger or frustration)." He goes on to say, "This specific character of bureaucracy means the complete eradication of love, hate, and all purely personal sentiments from administrative tasks. Put bluntly, this means the eradication of all sentiments that are irrational and incalculable."[10] Weber's vision appears to presage artificial intelligence, characterized by instrumental reason with no self-awareness and less emotion. However, to understand a bureaucracy (or AI), we must return to the human, not eliminate it, which is why an ethnographic approach, rather than an institutional approach, underpins this book. A fully developed bureaucracy is run and staffed by humans, whose lives and cultures and efforts matter, and whose foibles and fallibilities matter as well. Rather than eradicating sentiments, we must seek to understand them, even if they are constrained or warped by the system in place.

There are of course professional perils associated with writing about the very governmental bureaucracies that have hired me.[11] I sometimes challenge managerial representations and question the this-is-the-way-things-are point of view. I do so not simply to critique, even though informed critique has great value. I do so because alternative perspectives—ethnographic, historical, broadly comparative—provide the basis for improvement. I have found, however, that by challenging established managerial representations of public lands management, or by questioning the basis for this or that decision, I am typically met with one of three responses from my colleagues: mild interest, dismissive indifference, or puzzlement. On rare occasions, I have been advised by supervisors to keep my opinions to myself. My hope is that those who are interested, indifferent, or puzzled will see value in continuing to challenge the status quo. As I remind my colleagues, public lands management requires understanding "publics" and "lands," but not as separate categories of phenomena. These are thoroughly intertwined empirically and should be just as thoroughly intertwined managerially.[12]

Understanding such entwinement is challenging and also, apparently, dangerous, since it allows—indeed insists upon—a diversity of approaches to public lands management. Diversity itself tends to unsettle land managers who have been working in governmental bureaucracies the entirety of their adult lives. Anthropologist Mary Douglas, sounding a bit like Weber, refers to hierarchical social organization as exhibiting "a pathological tendency to try to control knowledge—for new knowledge is the biggest threat to its ordered ranking."[13] New ideas require effort to understand and implement and are often part of alternative discourses. Perhaps disruptive to the hierarchical status quo, they are easily dismissed as irrelevant. They can then be ignored simply because they discomfit. Preserving an ordered ranking, pathological or otherwise, often means precluding new knowledge from entering the hierarchical system. In this way, bureaucratic norms are maintained and conformity to them continuously established, even as some within the bureaucracy push against those norms and question that conformity.

Some of the chapters in this book are short. They are intended to provoke interest in coupled human and natural systems rather than analyze them. Other chapters are quite long, each exploring a topic rich enough to provide material for a separate book. But rather than writing lengthy monographs on individual topics—carefully tended rows of research, the monocrops of scholarship—I prefer to take an ecosystem approach to ideas. Ecosystems, as we all know, are messy, contingent, evolving, and unpredictable. They are not industrial and thoroughly regimented farms of corn, pine trees, or soybeans. The bureaucratic mind runs counter to the reality of natural and social worlds, wherever it develops knowledge ordered much like the geometry of tightly rowed pine trees and its celebration of nothing else growing there.

Since I work for the U.S. federal government, I should be explicit. The views herein are mine. They do not necessarily reflect agency point of view. In an age marked by all too many elected officials minimizing scientific influence in decision-making, I hold no illusions about anthropological scholarship influencing bureaucratic outcomes. But, as with the problem of objectivity, I've never been overly concerned with a lack of scholarly influence. Instead, I've simply tried to nudge the bureaucracy from within to improve environments and to forefront the humans who rely on and enjoy those environments. Recently, that nudge has turned into a push, as various environmental crises require a vigorous response.

Under George W. Bush, scientists, especially climate scientists, were often actively ignored. Under Barak Obama, the scientific community was embraced as a powerful force in understanding the world we live in. Under Donald Trump, the scientific community was not simply ignored, but openly castigated and directly undermined. My sense is that careful scientific thought will once again prevail, that the conceptual and governing sins of past administrations will be abandoned and then forgotten, and that our changing climate will demand involvement of our scientists as we try to sort out the natural and human consequences of our actions and our inactions.

By a return to science, I do not mean a return to Progressive-era conservation, with its belief in non-partisan, technocratic management.[14] Nor do I mean a return to the precepts of high modernism, which assumed social and environmental problems were amenable to technical solutions, organized by state power. All we needed to do was to measure, count, simplify, and rationalize those aspects of nature that were open to bureaucratic administration, and ignore the rest. If modern states and their constituent bureaucracies presume to rationalize nature and society, then crises function to upend those presumptions. Of course, especially in a crisis, we should remain clear-eyed about the many ways scientific effort can be bent to political purpose, and about the many ways political purpose can hide scientific understanding. It is easy enough to find examples of how politicians willfully fail to inform the public of hazards understood by the science of the day. A few chapters reflect on such willful public misinformation, which is not unique to our present circumstances.

The testing of atomic bombs in Nevada in the 1950s remains an important example, one which reminds us that politicians and the bureaucracies they govern can sacrifice fellow citizens for certain political purposes. At the end of the Cold War in the 1950s, fallout from open-air testing of nuclear weapons frequently drifted over small ranches and communities in southern Utah and Nevada and northern Arizona. Horses, cattle, and sheep were often exposed to the fallout; so were people, some of whom later died from cancers probably caused by the fallout.

The *precise* cause-and-effect relationships between fallout and cancers were not well-understood. However, the scientific community had by this time established links between radiation and various ailments such as malignant tumors, genetic defects, leukemia, and shortened lifespans. The Atomic Energy Commission was sufficiently concerned about the dangers to advise its employees involved with nuclear testing to take simple precautions with fallout. Stay indoors, they were told, and wash if exposed. Make sure to discard contaminated clothing. The science was clear enough: fallout is dangerous.

To the U.S. citizens who later became known as Downwinders, however, the Atomic Energy Commission had a different message. *Don't worry*, they were told, *there is no danger*. Many years later, a portion of Downwinders died from cancers more-likely-than-not caused by their exposure to fallout. The message to citizens, in clear opposition to the message to government employees, had been intended to mislead, and the government succeeded in misleading its citizens. Influential bureaucrats thought that open-air testing of nuclear weapons was more important than the safety of civilians, whom, if properly informed, may have found a way to curtail such testing. Powerless, once properly informed, they could have at least taken well-known precautions. Leave the area. Stay indoors. Wash. Toss out contaminated clothes.

"In the end," writes Philip Fradkin, "these people were betrayed by their government—the ultimate sin of a democracy." He goes on to say:

> It was a crime of betrayal perpetuated in the name of national security, a concept that may have been relevant to begin with but that shaded into bureaucratic intransigence as the Cold War waned, the tests went underground, and détente emerged ... At one end of the scale of injustice, this breach of faith could be viewed as an act of sustained stupidity, while at the other it resembled a perfidious act carried out by a government against its own people.[15]

Downwinders eventually filed a lawsuit. The U.S. federal district court judge who heard the case noted that it was

> concerned with the duty, if any, that the United States government had to tell its people, particularly those in proximity to the experiment site, what it knew or should have known about the dangers to them from the government's experiments with nuclear fission conducted above ground in the brushlands of Nevada during those critical years.

The case also focused on whether the United States could be legally excused "from being answerable to a comparatively few members of its population for injuries allegedly resulting from open air nuclear experiments conducted in response to such perceived dangers." The judge ruled against the federal government for its failure to inform citizens about the hazards of nuclear fallout.[16]

The federal government eventually won the legal battle and U.S. citizens, the Downwinders, lost. The U.S. Court of Appeals for the Tenth Circuit ruled that the so-called discretionary function trumped all other considerations in this case and overturned the district court's ruling. The discretionary function, a sort of throwback to the days of monarchies and sovereign immunity, says in essence that the government can do no wrong. It didn't matter, the appeals court opined, what level of bureaucratic functioning was involved, whether in development of policy (national security) or in operations carrying out policy (open-air testing of nuclear weapons). Discretion to act, even to the detriment of U.S. citizens, was a power held by the federal government.

The district court judge, however, saw things differently in this case, and reasoned that discretion to act required a measure of care: "responsible persons at the operational level of continental nuclear testing neglected an important, basic idea: *there is just nothing wrong with telling American people the truth.*"[17]

The same basic idea should apply to all governmental acts, from the construction of policy to the implementation of policy. Such truthfulness may seem quaint with a former U.S. president clocking some 30,573 lies and misleading claims in his four years in office.[18] But truthfulness is especially crucial for those social and environmental problems for which there are no good technical solutions, such as climate change. Scientific understanding may be incomplete, policymakers may be unable or unwilling to clearly define the relevant issues, stakeholders may articulate widely divergent opinions on preferred outcomes, and resource managers may have their hands tied by political appointees following the shifting winds of presidential elections. And in any modern polity, there are always differential relations of power—those who, in the social reality of a given moment, make decisions about our society, and those who don't. Such differential relations of power are as much part of proposed solutions to social and environmental problems as are scientific assessments. High modernism assumed science would undergird most administrative decisions. It is quite clear, however, that other social forces are at play.

One recent social force, referred to as "neoliberalism," may not succeed as a viable, long-term solution to environmental problems—despite the cheering from supporters of the capitalist economy, and despite some limited successes. Neoliberalism simplifies the natural world to market considerations. As with high modernism, it brackets and thereby ignores much of what makes natural and social worlds intriguingly complex and interconnected. Neoliberalism subordinates social and environmental policies to economic policies. It turns nature into a commodity subject to market forces. As importantly, neoliberalism is unable to accommodate cultural values that do not conform to the logic of markets.[19]

The lessons from high modernism apply to neoliberalism. Beware measures of value that become sources of value. Beware administrative processes that become the arbiter of what is real. Beware an imposed and narrowed field of bureaucratic vision.

The general point is that advocates of bureaucratic process or market rationality tend to wear the same kind of organizational blinders. They see the world simplified. Step far enough outside of bureaucratic process and you no longer are visible to that process—except, perhaps, as a threat. Assert values that are at odds with the capitalist economy or are critical of the notion of a self-interested actor at the heart of all consumption, and you will be marginalized.

Yet we all know the world is more complex and interesting, and indeed different, than either form of simplification can envision. Hence, in the broad scheme, we need better science and much more of it.

As a collective, self-critical enterprise, scientific research may call into question the assumptions animating bureaucratic process and may better illuminate the environmental effects of market economies. This is also why, again in the broad scheme, some elected officials—indeed, all too many citizens—are disdainful of science. They do not want their assumptions about the world and its functioning to be questioned, nor do they want the environmental effects of market economies to be scrutinized. Some elected officials go so far as to require scientists working for them to vet their research through political appointees prior to its publication. They can thereby squelch whatever they disagree with, for whatever ideological reasons. Exercising the differential relations of power of a bureaucracy, they tend to wield the ham fist.[20]

Some large bureaucracies—the U.S. Department of Agriculture employees some 105,000 people, for example—contain suborganizations that are tucked away, and within them scientists continue to do productive work. This is not so much the "deep state," so named to make us fearful, as it is the "deep-thinking state." One curiosity about the narrowed field of bureaucratic vision is that high-level bureaucrats, especially political appointees, often fail to see the very organization they work within or administer. Tucked away in the corners of many governmental bureaucracies, scientists persist. One small example from my own world concerns the Department of Agriculture, which under the Trump administration discouraged the use of the phrase "climate change." Nonetheless, it maintained a climate change program located in the office of the chief economist.

The climate change program focuses on the "implications of climate change on agriculture, forests, grazing lands, and rural communities."[21] It produces scientific reports intended to inform and direct policy. Its scientists, perhaps discouraged by political direction, continue nevertheless to analyze climate change and its effects on natural and human environments—even as policymakers find ways to ignore those very reports and to hide them from the public.[22]

By describing governmental bureaucracies in stark terms, I too indulge in the very simplification I attempt to avoid. I recognize the tendency and hope the following chapters provide sufficient response. I also recognize that bureaucracies have come to dominate much of our social existence. They are not primarily governmental. The other kinds of bureaucratic structures play outsized roles in our lives. Indeed, they play outsized roles in global affairs. The International Monetary Fund, the World Trade Organization, and the World Bank, for example, are bureaucratically organized, as are financial mega-firms such as Goldman-Sachs, American International Group, and (now defunct) Lehman Brothers. To imagine that any of these sorts of organizations stand entirely outside of governmental oversight is of course absurd. The notion of a "free market" is ideological, not empirical.[23]

The governance question is what sort of oversight conducted by what sort of governmental agency wielding what sort of power should ensure market forces don't result in degraded environments. The related question is this: should governments corral financial firms interested in their own profit, not in the public good? I think the answer is yes. Of course, these sorts of financial firms are "planetary bureaucracies," and there are no corresponding planet-wide governance forms to keep them in check.[24] Still, one responsibility of government within a democracy is to make sure economies function for the larger good and not solely for a small class of "investors." And yet the growth and increasing bureaucratic complexity of financial mega-firms and their too-big-to-fail power appears to mostly benefit a small segment of our society, rather than the whole. Moreover, the juggernaut of modern financial mega-firms, including their entanglements with government, clearly influences environments, environmental policies, and public lands management. In my own simplification, I've barely touched on those influences in these pages. More work remains to be done.

Anthropologist David Graeber suggests we may be in the "age of total bureaucratization."[25] In such an age, public and private bureaucratic structures overlap or are fused in all sorts of ways. Certain overlaps are of questionable provenance, as when former U.S. Vice President Dick Cheney met with Halliburton executives to plot out U.S. energy policy—a meeting that remains hidden and mysterious, both in content and consequence (talk about the deep state). In such circumstances, governmental oversight becomes governmental complicity.[26] Other overlaps are understandable and aboveboard, as when the Environmental Protection Agency enforces laws prohibiting certain forms of air pollution from coal-burning power plants, or when the Bureau of Land Management works with ranchers to ensure their sheep or cattle do not degrade public lands, even as some people in rural communities see federal administration of public lands as a disturbingly alien form of power.

Scientific reports, economic assessments, environmental impact statements, categorical exclusions, National Environmental Policy Act, National Historic Preservation Act, Council on Environmental Quality regulations, State Historic Preservation Offices, Tribal Historic Preservation Offices, permits, licenses, inholdings, easements, boundaries, jurisdiction, findings of no significant impact (FONSI), adverse effect, no adverse effect, scoping, record of decision (ROD), instruction memorandums, resource values, fees, fines, public hearings, Memorandum of Agreement, Memorandum of Understanding, environmental justice, endangered species, threatened species, cultural resources, briefing papers, news releases, Freedom of Information Act (FOIA), private meetings among elected officials, public meetings of resource advisory committees, federal register notices, Section 106, lawsuits, judgements, appeals, political posturing, chains of command, and a seemingly endless parade of forms to fill out—the stuff of bureaucratic functioning—all are part of the process of oversight.

The language of bureaucracy is like mud, turbid and unpalatable. Little wonder such language and the accompanying bureaucratic functioning appear alien to ranchers herding cows. They probably appear alien to most people, outside of the certified experts speaking the jargon-filled language and involved in the bureaucratic functioning itself. Esoteric, specialized, and largely inaccessible bureaucratic knowledge may well form the basis of administrative power generally, as Weber foresaw.[27]

In an age of total bureaucratization, we may also be in an age that defies bureaucratization. Not only in large public displays, such as the 2014 Cliven Bundy standoff, which included an armed confrontation between his supporters and federal officers over his failure to pay grazing fees for his use of public lands in Nevada; or the 2016 occupation of the Malheur National Wildlife Refuge in Oregon, led by Ammon Bundy, which also occasioned an armed confrontation, and was intended to underscore the issue of state versus federal power. Such public displays did little to change bureaucratic structures. In many ways, they occasioned more bureaucratic effort, with the involvement of the federal judiciary and the attendant paperwork required.[28]

I have in mind a very different sort of defiance. Perhaps "defiance" is the wrong term to use.[29] Some people work the bureaucratic system by staying hidden from that system. To put it the other way around, bureaucratic systems see what they are set up to see, and little else. Anthropologist Anna Tsing, in her wonderfully engaging book, *The Mushroom at the End of the World*, does not directly talk about bureaucracies. She describes, rather, "the possibility of life in capitalist ruins."[30] Among other things, she is interested in what happened when Pacific Northwest forests in the United States were denuded of old-growth trees, the result of governmental mismanagement and decades-long commercial efforts. On public lands in the Pacific Northwest, fires were suppressed, ponderosa pine were clear cut, huge profits were made, and then, by the late 1980s, the timber economy collapsed. Despite the efforts of the Forest Service to regrow ponderosa pine, what grew in to replace them in those ruined industrial forests were trees of lesser economic value: lodgepole pine and fir trees. As it turned out, ponderosa pine depended upon periodic fires to flourish, whereas lodgepole pine thrives in fire-suppressed landscapes.[31]

Lodgepole pine—spindly, crowded, prone to burn—provides the perfect companion tree for a kind of fungus called matsutake, which has a market in another part of the world. Matsutake takes about 40 to 50 years to fruit in fire-suppressed lodgepole pine forests. A delicacy in Japan, valued at up to $700 a pound during the 1980s, it commands a high price to this day. Just as the debates in the United States over old-growth forests, spotted owls, and the collapsed timber economy were in full swing, word began to spread about matsutake growing in Oregon. But word spread in such a way that few people noticed, at least initially.

The Chernobyl nuclear disaster in 1986 provided an unexpected boost to matsutake harvesters in Oregon. The fallout from that disaster contaminated mushrooms in Europe, rendering that source of matsutake unmarketable. Various wars also provided an unexpected source of self-directed, independent labor, as Mein, Lao, Khmer, and Hmong—part of a Southeast Asian diaspora—found work harvesting mushrooms in Pacific Northwest national forests, joined by other people seeking a measure of economic independence. The work was hard, not particularly lucrative for the harvesters, and required an intimate knowledge of forests and fungi. Part of an informal network of harvesters, buyers, and exporters, these people stayed mostly invisible to the bureaucracies involved with timber production, conservation, or public lands management. Eventually, the Forest Service caught on and took various steps to regulate matsutake harvesters and matsutake harvests, the value of which had come to exceed the value of timber being cut on national forests in the Pacific Northwest.

Both the Bundy standoffs and matsutake harvesters illustrate one of Scott's points about high modernism and state power, cited earlier: the frustrated goal of the modern state is to reduce the chaos and transform the world into something resembling the administrative grid of its own observation. Wild gardens then become industrial plantations. Not all humans agree to conform to that goal however, setting up further conflicts over proper governance, power, and individual agency.[32] And in such conflicts, we see different knowledge systems vying for a public hearing and bureaucracies trying to understand, accommodate, and thwart them.

Recent reports show that the last two human generations have coincided with a 60 percent loss of wildlife and a 70 percent loss of insects worldwide.[33] Since 1970, humanity has witnessed and indeed caused an overall and disturbing decline in wildlife and insect populations. If that pace continues, we will soon enough know the worth of freshwater, saltwater, and terrestrial wildlife, especially if climate change intensifies and the pace of loss accelerates.[34] Our best hope for our own survival is to maintain healthy natural systems at the same time we improve our social systems. In a small way, this book holds out that hope, even as it critiques the nature of bureaucracy, and questions the bureaucratization of nature.

Notes

1 Herbert Kauffman, *The Forest Ranger: A Study in Administrative Behavior* (Washington, DC: Resources for the Future, 2006), 37.
2 Weber's quotes are from Tony Waters and Dagmar Waters, eds. and trans., *Weber's Rationalism and Modern Society: New Translations on Politics, Bureaucracy, and Social Stratification* (New York, NY: Palgrave Macmillan, 2015), 76, 95, 97, 116, 115.
3 James C. Scott, *Seeing Like a State: How Certain Schemes to Improve the Human Condition Have Failed* (New Haven, CT: Yale University Press, 1998), 80–81.
4 Peter M. Vitousek, Harold A. Mooney, Jane Lubchenco, and Jerry M. Melillo, "Human Domination of Earth's Ecosystems," *Science* 277:5325 (July 25, 1997), 494–499.
5 Joanne Bauer, ed. *Forging Environmentalism: Justice, Livelihood, and Contested Environments* (Armonk, NY: M.E. Sharpe, 2006). This effort was sponsored by the Carnegie Council for Ethics in International Affairs. www.carnegiecouncil.org/education/006/forging_e nvironmentalism.
6 David Jenkins, "Atlantic Salmon, Endangered Species, and the Failure of Environmental Policy," *Comparative Studies in Society and History* 45 (October 2003), and "When the Well's Dry: Water and the Promise of Sustainability in the American Southwest," *Environment and History* 15 (November 2009).
7 David Jenkins, "Impacts of Neoliberal Policies on Non-Market Fishing Economies on the Yukon River, Alaska," *Marine Policy* 61 (2016).

8 See William Shands, "The Lands Nobody Wanted: The Legacy of the Eastern National Forests," in Harold Steen, ed., *The Origins of the National Forests* (Durham, NC: Duke University Press, 1992); David E. Conrad, *The Land We Cared For…A History of the Forest Service's Eastern Region* (Washington, DC: U.S. Department of Agriculture, 1997).

9 See David Jenkins, "Anthropology, Mathematics, and Per Hage's Contribution to Kinship Theory," in *Kinship, Language, and Prehistory: Per Hage and the Renaissance in Kinship Studies*, Doug Jones and Bojka Milicic, eds. (Salt Lake City, UT: University of Utah Press, 2011), and "The Ethnography of the Self: Anthropologists' Autobiographies," in *The Character of Human Institutions: Robin Fox and the Rise of Biosocial Science*, Michael Egan, ed. (New Brunswick, NJ: Transaction Publishers, 2014).

10 Waters and Waters, eds., *Weber's Rationalism and Modern Society*, 97.

11 For examples of how government-employed scientists with reasoned views contrary to the prevailing bureaucratic orthodoxy were hounded out of their jobs, often in petty, politically motivated ways, see Todd Wilkinson, *Science Under Siege: The Politicians' War on Nature and Truth* (Boulder, CO: Johnson Books, 1998).

12 See Rosemary E. Ommer, *Coasts Under Stress: Restructuring and Social-Ecological Health* (Montreal: McGill-Queen's University Press, 2007), for a detailed, comparative study.

13 Mary Douglas, *Natural Symbols: Explorations in Cosmology* (London: Routledge, 2003), xxv.

14 As described in Samuel P. Hays, *Conservation and the Gospel of Efficiency: The Progressive Conservation Movement, 1890-1920* (Cambridge: Harvard University Press, 1959).

15 Philip L. Fradkin, *Fallout: An American Nuclear Tragedy* (Boulder, CO: Johnson Press, 2004; first published in 1989 by the University of Arizona Press), 25. See also Morgan Knibbe's documentary film, "The Atomic Soldiers" (2019), for firsthand accounts as veterans break their government-imposed silence on the experience of being exposed to nuclear weapons testing.

16 *Irene Allen et al. v. The United States of America*, Memorandum Opinion, May 10, 1984. I cite my father, District Court Judge Bruce S. Jenkins. He continues to hear cases, quite remarkable for a man approaching 95 years old.

17 Emphasis in the original. *Allen v. U.S.*, 317.

18 Glenn Kessler, "Trump made 30,573 false or misleading claims as president. Nearly half came in his final year," *The Washington Post* (January 23, 2021). Glenn Kessler, Salvador Rizzo, and Meg Kelly, "President Trump Has Made More Than 20,000 False or Misleading Claims," *The Washington Post* (July 13, 2020). Glenn Kessler, Salvador Rizzo, and Meg Kelly, "President Trump Has Made 15,413 False or Misleading Claims over 1,055 Days," *The Washington Post* (December 16, 2019). Glenn Kessler, Salvador Rizzo, and Meg Kelly, "President Trump Has Made More than 10,000 False or Misleading Claims," *The Washington Post* (April 29, 2019). Glenn Kessler, Salvador Rizzo, and Meg Kelly, "President Trump Made 8,158 False or Misleading Claims in His First Two Years," *The Washington Post* (January 21, 2019).

19 See Wendy Brown, *Undoing the Demos: Neoliberalism's Stealth Revolution* (New York, NY: Zone Books, 2015), and David Harvey, *A Brief History of Neoliberalism* (Oxford: Oxford University Press, 2007). Also of interest is Anand Giridharadas, *Winners Take All: The Elite Charade of Changing the World* (New York, NY: Alfred A. Knopf, 2018).

20 See Michael Lewis, *The Fifth Risk* (New York, NY: W.W. Norton & Company, 2018) for an account of the failures of leadership transition for the Trump administration, and especially for accounts of a simple lack of interest in science and a related lack of interest in being educated about how government works and the benefits it provides. As senior advisor to the president and son-in-law Jared Kushner said, apparently without a hint of irony or self-awareness, "We've read enough books." Sarah Vowell, *The New York Times* (August 8, 2017).

21 www.usda.gov/oce/climate_change/index.htm.

22 For a handful of examples among many, see Coral Davenport, "Trump Administration's Strategy on Climate: Try to Bury Its Own Scientific Report," *The New York Times* (November 25, 2018), Eric Lipton, "Interior Nominee Intervened to Block Report on Endangered Species," *The New York Times* (March 26, 2019), Juliet Eilperin and Brady Dennis, "New EPA Document Tells Communities to Brace for Climate Change Impacts," *The Washington Post* (April 28, 2019), Coral Davenport and Mark Lander, "Trump Administration Hardens Its Attack on Climate Science," *The New York Times* (May 27, 2019), Gretchen T. Goldman, "Trump's Plan Would Make Government Stupid," *Nature* 570 (June 20, 2019), Helena Bottemiller Evich, "Agriculture Department Buries Studies Showing Dangers of Climate Change," *Politico* (June 23, 2019), Helena Bottemiller Evich, "Trump's USDA Buried Sweeping Climate Change Response Plan," *Politico* (July 18, 2019), Hiroko Tabuchi, "A Trump Insider Embeds Climate Denial in Climate Research," *The New York Times* (March 2, 2020), and Christopher Flavelle, "How Trump Tried, but Largely Failed, to Derail America's Top Climate Report," *The New York Times* (January 1, 2021). These examples keep multiplying. After a highly successful effort to improve school nutrition which benefited 30 million schoolchildren, led by the former Secretary of Agriculture in the Obama administration, Secretary Sonny Perdue announced the USDA would not require schools to meet high nutritional standards, and instead lowered those standards, which increased profits of those companies supplying meals, to the detriment of children who ate them. See Lewis, *The Fifth Risk*, 104–107.

23 This fact has been well known since at least Karl Polanyi, *The Great Transformation: The Political and Economic Origins of Our Time* (New York, NY: Farrar and Rinehart, 1944). See also David Graeber, *Debt: The First 5,000 Years* (Brooklyn, NY: Melville House, 2011).

24 David Graeber and David Wengrow, *The Dawn of Everything: A New History of Humanity* (New York, NY: Farrar, Straus and Giroux, 2021), 431.

25 David Graeber, *The Utopia of Rules: On Technology, Stupidity, and the Secret Joys of Bureaucracy* (Brooklyn, NY: Melville House, 2015). See also Michael Herzfeld, *The Social Production of Indifference: Exploring the Symbolic Roots of Western Bureaucracy* (Chicago, IL: University of Chicago Press, 1992).

26 For a history of such complicity, see Richard White, *Railroaded: The Transcontinentals and the Making of Modern America* (New York, NY: W.W. Norton & Company, 2011). It is instructive to read White's *Railroaded* in conjunction with Jane Mayer's *Dark Money: The Hidden History of the Billionaires Behind the Rise of the Radical Right* (New York, NY: Anchor Books, 2017). White demonstrates in detail how nineteenth-century railroad tycoons—often bumbling and ill-informed—complicit with government officials, made vast personal fortunes at the expense of the public, which underwrote their mostly failed railroad schemes. Mayer, in similar detail, demonstrates how the Koch brothers in the modern era are trying to do essentially the same thing: assemble personal fortunes based on public resources. In both circumstances, complicit governmental bureaucracies are key players.

27 In the financial sector, the complex financial products flowing from financial bureaucracies that contributed to the 2008 economic collapse and to its long-term consequences continue to baffle: commodity derivatives, mortgage-backed securities, hedge funds (the "shadow banking system"), tranches, credit default swaps, obligation derivatives, adjustable rate mortgages, collateralized dept obligations, triple-A ratings, subprime mortgages, Fannie Mae, Freddie Mac—financial products and quasi-governmental organizations that were too complicated for either the originators or ordinary mom-and-pop participants to fathom. See Michael Lewis, *The Big Short: Inside the Doomsday Machine* (New York, NY: W.W. Norton & Company, 2010).

28 James R. Skillen, *This Land Is My Land: Rebellion in the West* (New York, NY: Oxford University Press, 2020), provides a useful overview. See also Christi Turner, "Timeline: The BLM vs. Cliven Bundy," *The High Country News* (May 12, 2014), Tay Wiles, "Malheur Occupation, Explained," *The High Country News* (January 4, 2016), as well as robust associated reporting from HCN. The American West is both a product of and thoroughly dependent upon the federal government, even as rural folk and various state politicians fail to acknowledge that historical fact. See Richard White, *'It's Your Misfortune and None of My Own': A New History of the American West* (Norman, OK: University of Oklahoma Press, 1991).

29 See, however, James C. Scott, *Weapons of the Weak: Everyday Forms of Peasant Resistance* (New Haven, CT: Yale University Press, 1985). With increasing economic disparity, such weapons of resistance may well emerge in modern guise.

30 Anna Lowenhaupt Tsing, *The Mushroom at the End of the World: On the Possibility of Life in Capitalist Ruins* (Princeton, NJ: Princeton University Press, 2015).

31 See Nancy Langston, *Forest Dreams, Forest Nightmares: The Paradox of Old Growth in the Inland West* (Weyerhaeuser Environmental Books, 1996).

32 See John F. Devlin, ed., *Social Movements Contesting Natural Resource Development* (New York, NY: Routledge, 2020).

33 See The Living Planet Report 2018 by the Zoological Society of London and the World Wildlife Fund. www.zsl.org/global-biodiversity-monitoring/indicators-and-assessments-unit/living-planet-index. Human influence on mammalian ecology is quite old, even as it has picked up speed. See Felisa A. Smith, Rosemary E. Elliott Smith, S. Kathleen Lyons, Jonathan L. Payne, and Amelia Villaseñor, "The Accelerating Influence of Humans on Mammalian Macroecological Patterns over the Late Quaternary," *Quaternary Science Reviews*, v 211 (May 2019). On insects, see Dave Goulson, *The Silent Earth: Averting the Insect Apocalypse* (New York, NY: HarperCollins, 2021).

34 See J.R. McNeill and Peter Engelke, *The Great Acceleration: An Environmental History of the Anthropocene Since 1945* (Cambridge: Harvard University Press, 2014). For specific and accessible examples of the interaction of industry, government, and environment, see Dan Egan, *The Death and Life of the Great Lakes* (New York, NY: W.W. Norton & Company, 2017); Michael Pollan, *The Omnivore's Dilemma: A Natural History of Four Meals* (New York, NY: The Penguin Press, 2006); Meghan L. O'Sullivan, *Windfall: How the New Energy Abundance Upends Global Politics and Strengthens America's Power* (New York, NY: Simon & Shuster, 2017). See also Ingrid J. Visseren-Hamakers and Marcel T.J. Kok, eds., Transforming Biodiversity Governance (Cambridge: Cambridge University Press, 2022). The literature is large; these are a few places to start.

PART I

The Bureaucracy of Nature

1

AGAINST EFFICIENCY

Why We Cut Trees (And What Happens When We Do)

One muggy summer day in 2017, I pitched the importance of bureaucratic ineffi-ciency to an audience of 40 or so U.S. Forest Service leaders. We had assembled in Ohio near the Wayne National Forest to talk about trees. The Forest Service had been charged by the Trump administration with increasing its national timber pro-duction by 15 percent. However, existing bureaucratic timber harvesting processes were seen as getting in the way of that goal. The call was then made to make those processes more efficient and thereby increase the sale of timber on the federal public lands we managed. Policy yields implementation. Simple. Political cause, in this model, generates empirical results. We even had a wonderful phrase, turned inev-itably into an acronym, to guide our efforts: Environmental Analysis and Decision Making—EADM—pronounced "edam," like the cheese.

The initial bureaucratic battle was won. Those who devised policy were delighted. But the rest, the on-the-ground implementation, the cutting of all those trees, remained to be seen. The primary bureaucratic obstacle to the plan appeared to be the National Environmental Policy Act (NEPA), passed during the Nixon administration in 1969 and enacted on January 1, 1970. This is the law that requires environmental assessments or environmental impact statements for *proposed* fed-eral actions, such as timber harvests, which have environmental consequences. Its purpose is to analyze and then avoid or mitigate negative effects of *actual* timber harvests. A second bureaucratic obstacle was another act, the National Historic Preservation Act (NHPA), passed in 1966, which some in the timber world envisioned as improperly slowing timber harvests. NHPA requires surveys of cul-tural significance. It insists we chart historic and prehistoric evidence of human occupation prior to any disturbance of federal land. As with NEPA, NHPA requires us to avoid or mitigate negative effects of timber harvests.

In the Ohio meeting, I pointed out that both the National Environmental Policy Act and the National Historic Preservation Act were acts of inefficiency.

DOI: 10.4324/9781003297444-3

They required us to stop and think. They required us to act only if we had carefully thought about the consequences of our actions. I argued that the American public, through its Congress, wanted us as federal lands managers to take the time needed to adequately analyze our actions, and that becoming "efficient" in many ways curtailed our thoughtfulness. Hence the need for *inefficiency*. It was, in my argument, the consequence of thoughtfulness.

My colleagues responded with a circus of complaints. Without doubt, they emphatically said, we needed to be more efficient. There was always room to improve our bureaucratic processes. Let's not get in our own way and be our own worst enemy by remaining inefficient. In reply, I remarked that we had turned "efficiency" into a fetish and, as with all fetishes, we overlooked important details. What were we hiding with our quest to be efficient? What powers were involved? What was lost? Who, after all, got to make the decision and at what level of the organization? My effort that day was to gently question the collective wisdom.

Maybe I should not have said "fetish." But the point I raised still bears some scrutiny, precisely because of the multiple negative reactions to it. Apart from using anthropological jargon, I'd violated a norm of some sort. I'd questioned what was held to be obviously true. I'd come out against efficiency.

I can imagine readers banging the side of their head in disbelief. Bureaucracies are famously inefficient (contrary to Max Weber's view of perfected bureaucratic efficiency). The rules and regulations, and also the insistence on rules and regulations in the face of contrary evidence, seem to characterize all bureaucracies, not just those involved in public lands management. And I'm aware of the large and growing scholarly literature critical of bureaucratic force and culture, a literature that demonstrates the failures of bureaucratic thought. Indeed, I've contributed to that literature. Bureaucracies, and those who become enculturated within them, can be stridently implacable and goofily officious. I often bang my own head in disbelief.

But let me explain my argument against efficiency. It winds through a century of forests, wildfires, and organizational history. It follows "efficiency" being transformed from a technical measure to a moral attribute. It dips into soil science and samples policies of atmosphere. Finally, my analysis finds value in inefficiency, which I define simply as taking the time to think. Inefficiency comes with the territory.

Those of us who grew up during the 1960s environmental movement know the importance of NEPA. We need time for our scientists to assess everything from endangered species to watersheds, from prehistoric humans to ecological consequences of building roads. With any disturbance to national forests, we need to avoid destroying species and extinguishing evidence of past cultures. Congress had given the U.S. Forest Service direction on those very issues. Follow the science, Congress said, and only then proceed if warranted or to the extent warranted.

Built into the process was ample opportunity for citizens to help us stay on the right path. Everything about the process took time and effort. It required thoughtfulness from many perspectives, scientific, indigenous, commonsensical, oppositional, supportive, economic, perhaps even political. If a social group or an individual wanted to influence public land management decisions, then there was plenty of opportunity to do so. The result for any particular timber harvest was never guaranteed in advance of adequate analysis. This was the modern ideal.

This ideal, especially citizen involvement, was needed because of prior failures of timber management. Many of those failures were the product of Progressive Era approaches to managing public forests in pursuit of orderly timber harvests. Other failures flowed from market-dominated policies. Still others resulted from a focus on timber, rather than on wildlife or ecosystems. I feared we were once again headed in the wrong direction as we focused solely on trees and their efficient harvest. To understand my concern, we need to begin with the efforts of the first Chief Forester, Gifford Pinchot, and with the origins of the Forest Service.

A Utilitarian Vision

In 1898, President McKinley appointed Gifford Pinchot as Chief of the U.S. Division of Forestry. Seven years later, he became the first Chief Forester of the U.S. Forest Service. Trained in German forest management, Pinchot instilled in the U.S. Forest Service a sense of mission. He promoted rational, scientific, and efficient management of timber harvests on public lands. He insisted that timber professionals and not politicians should run the program. He believed whenever politicians were involved efficiency would disappear and processes would become corrupted. "Forestry is Tree Farming," Pinchot proclaimed. "To grow trees as a crop is Forestry."[1]

As he extolled the virtues of scientific forestry, Pinchot also managed to keep out other concerns, such as those around big game management on public lands and those associated with the nascent late nineteenth- and early twentieth-century cultural movement favoring a "spiritual" relationship with nature. John Muir, founder of the Sierra Club, stood as Pinchot's ideational opposite during this time, with Ralph Waldo Emerson and Henry David Thoreau as Muir's kindred spirits.[2]

Muir and Pinchot were friendly, at least initially, and spent many weeks together tramping through western forests. They talked about nature and mused about humanity's responsibilities for the natural world. Muir, the older of the two, remained intent of preserving as much of nature as possible, for what he saw as nature's intrinsic value. "Thousands of tired, nerve-shaken, over-civilized people," Muir wrote in 1898,

> are beginning to find out that going to the mountains is going home; that wildness is a necessity; and that mountain parks and reservations are useful not only as fountains of timber and irrigating rivers, but as fountains of life.[3]

Pinchot also recognized the value of nature. He was keen on holding corporate interests at bay, in large part because he saw such interests as self-serving and destructive of the environment, even as he insisted that utilitarian values form the basis of conservation. His oft-repeated refrain, "the greatest good of the greatest number in the long run," contained a strong materialist orientation measurable in board feet, not in the necessity of wildness. This phrase, coined by Jeremy Bentham, was later adopted by John Stuart Mill. Pinchot may have added "in the long run," in a 1905 letter from the Secretary of Agriculture to Pinchot describing Pinchot's duties—a letter Pinchot himself wrote. I've heard many in the Forest Service use this phrase, but when asked they seem not to know its history. They tend to believe, in a Forest Service received wisdom sort of way, Pinchot's self-promoting claim that he coined the phrase himself.[4]

Pinchot's greatest good utilitarian vision—to ensure forests remained in the public domain but as "working forests"—was eventually realized, with the crucial involvement of another friend and outdoor companion, Theodore Roosevelt. Elevated to the presidency after the assassination of President McKinley in 1901, Roosevelt helped Pinchot expand the national forest system from 60 *forest reserves* containing 56 million acres in 1905, to 150 *national forests* containing 172 million acres in 1910. This expansion (and name change) proceeded despite fierce opposition from influential senators bent on the immediate plundering of public wealth for personal profit. Montana Senator William A. Clark, for example, expressed sentiments which resembled those of railroad tycoons, whose companies had received over 131 million acres in land grants from the United States.[5] Clark, the richest man in Montana, remarked in 1903, "Those who succeed us can well take care of themselves."[6] Pinchot, Roosevelt, and Muir found this attitude reprehensible.

Pinchot was faced with an immense challenge. How could a small, underfunded agency—the nascent Forest Service—possibly manage a huge swath of national forests in diverse environments? The task was even more daunting because Roosevelt had outmaneuvered a reluctant Congress in the expansion of public lands, and many legislators resented being snookered by the president. One of Pinchot's solutions was to hire foresters, most of whom were trained at Yale University in a new department endowed with funds from Pinchot's well-to-do family. These foresters were then sent out to the new forests to begin the work of management, more-or-less independent of the whims of Congress. Pinchot made sure to hire like-minded men, who in turn responded with loyalty to him and his cause, as they set about the unprecedented task of large-scale scientific forest management.

Pinchot's utilitarian ethos of efficient, scientific forestry held sway for many decades, even after he was fired for political reasons during the Taft administration—indeed, it continues to hold sway. "The crux of the gospel of efficiency," writes historian Samuel P. Hays, "lay in a rational and scientific method of making basic technological decisions through a single, central authority."[7] For a time, Pinchot, with Roosevelt's support, was that authority. What, he wondered, was the economic value of nature, and how can government best catalogue and exploit such value? Elevating concepts of efficiency, functionality, simplification, and monetary worth to high status, Pinchot effectively pushed aside other values in the larger quest to sustainably harvest trees, aware future generations depended on his efforts.

From Pinchot's ouster in 1910 until 1920, Henry S. Graves served as the second Chief Forester. He also couched most forestry issues in economic terms. In his 1911 book, *Principles of Handling Woodlands*, Graves directly opposes clearing land for agriculture—for which trees were discarded as a nuisance—with what he calls "lumbering for use," which generated income. Lumbering for use had obvious economic advantages for private industry, which moved from forest to forest without regard for environmental consequences. Short-term timber profits superseded long-term sustainability, with very few exceptions. For industry, regrowth of forests was at best an afterthought. Graves, following Pinchot's lead, wished to change that orientation, and public lands provided the opportunity for the transformation.

Lumbering for use, however, was not without its own contradictions. Graves points out several. "It was soon after lumbering for use began," he writes,

> that forest fires became a common occurrence, and these increased in number and severity, burning over the majority of lumbered lands, and usually at the same time enormous areas which had not been cut, and destroying millions of dollars worth of timber.

Fifty million acres since 1870 had burned, and such burning destroyed timber valued at $50 million annually. It was clear that lumbering for use resulted in fires, which further consumed new growth in the cutover acres and burned up old growth in the adjacent uncut forests.[8]

For Graves, the "underlying idea of forestry is continuity of use." Public forests have been "set aside, to be managed for the permanent benefit of the public."[9] The benefits of public forests, however, were not to flow primarily to industry. They were to flow in perpetuity to the public. How? Graves offers a general management plan for cutting trees, which is site and species-specific, geared toward continuity, and sensitive to market conditions.

Briefly, his management plan calls for either cutting all trees in a stand—the Clear-Cutting, Shelterwood, and Coppice Systems—or only cutting select trees in order to preserve stand integrity—the Selection System. The Clearcutting System, in his view, had ten types, from clearing the entirety of a stand and restocking with seedlings, to clearing a stand with the exception of certain trees left for seed, to clearing a stand sequentially in strips or patches. The Shelterwood System, by contrast, slowly removes a given stand through sequential thinning, resulting in its eventual self-replacement. The Coppice System clears all hardwoods from the landscape and then relies on regeneration from the stumps of trees to rebuild a forest. In the Shelterwood and Coppice Systems, reseeding or planting seedlings were generally not required.

Despite emerging successes managing national forests according to principles of efficient, scientific forest management, problems appeared, including an overreliance on scientific rationality as defined and administered by a single, central authority. Two mistakes stand out, both accentuated by the third Chief Forester, William

B. Greeley, who was in charge of the Forest Service from 1920 to 1928. The first mistake was allowing extreme commercial clear-cutting on national forests. Under Greeley's watch, the timber industry succeeded in capturing regulatory control of the national forests.[10] Grave's "Selection System," favored by Pinchot, dropped out, and "Clear-Cutting Systems," favored by Greeley, took over. In this way, Pinchot's vision of working forests transformed into Greeley's reality of worked-over forests. Greeley summed up the contrast in these terms:

> Gifford Pinchot and I looked at the economic side of the forest picture through different glasses. He saw an industry so blindly wedded to fast and destructive exploitation that it would not change. I saw a forest economy overburdened with cheap raw material. Mr. Pinchot saw a willful industry. I saw a sick industry. G.P. disagreed sharply with the 1916 Forest Service report on lumber, in which I tried to give a factual picture of the underlying economic troubles. He called it a "whitewash of destructive lumbering."[11]

Herb Block's political cartoon from a 1929 Chicago newspaper captures the consequences (Figure 1.1).

The second mistake was of extreme fire suppression, which continued throughout much of the twentieth century. Influential timber professionals in both government and industry insisted on suppressing all fires, natural and human caused, which had the effect of also suppressing the science showing the ecological importance of fire-adapted landscapes. Although Pinchot and Graves were advocates of fire suppression, it was Greeley who pushed hard to actively suppress all forest fires. Along the way, he pushed just as hard to ignore any science contrary to his position. He also downplayed cultural practices associated with setting fire to landscapes.[12] "Throughout the South," he wrote, "man-set fires were well nigh universal." He goes on to say:

> They were part of the accepted order of things. Farmers and loggers and turpentine orchardists all fired the woods. They set fires to "green up" the forage, to uncover oak and beech mast for their razorback hogs, to clean out litter before boxing pines for resin, to open up the brush for better hunting, to get rid of cattle ticks or chiggers or snakes, or just because the woods had always been burned now and then.[13]

Notwithstanding such cultural practices, Greeley was convinced fire was the enemy. "[F]ire prevention is the No. 1 job of the American forester," he wrote. "[T]he first and greatest commandment of American forestry is to keep fire out of the woods." He added, "Eternal vigilance still is the price of forestry."[14]

There is historical irony in Greeley's position. He established in 1916 the scientific research branch of the Forest Service, much of which was dedicated to the study of fire.

"THIS IS THE FOREST PRIMEVAL——"

FIGURE 1.1 "This Is the Forest Primeval"

Source: *Chicago Daily News*, April 24, 1929. Copyright Herb Block Foundation. Used with permission.

The Enemy

In the early days of the Forest Service, fire prevention and fire suppression were at best haphazard. Too few rangers were responsible for putting out too many fires. Roads and trails were nonexistent in most of the new national forests, which made quick access to fires a problem. In some forests, railroads continued to build lines and operate without oversight and were the source of many forest fires. In one instance in Idaho, a ranger was assigned to follow rail cars and put out any consequent fires; however, the task proved overwhelming for one man on a velocipede. The fire season in 1910, which at its height burned three million acres in two days in Idaho and Montana, killed 85 people. Costing the Forest Service more than one million dollars, that catastrophic fire was a turning point for forest management.

In 1910, 29-year-old William Greeley was in charge of some 30 million acres in Montana, Idaho, and South Dakota. In this vast landscape, he oversaw 160 rangers, each of whom was responsible for about 190,000 acres of National Forest. He worked hard to develop cooperative agreements with timber associations, railroad companies, and states to patrol for and extinguish fires. It was a formidable assignment, made more so with the fires which erupted that year. Greeley (who ten years later became Chief of the Forest Service) had warned all his rangers to be especially watchful and reminded them that summer humidity had "dropped to the level of the Mojave Desert." The forests were bone-dry, and Greeley worried. In late July, a violent electrical storm roared through the region, igniting some 1,000 fires in 22 national forests. Greeley and his rangers managed to quickly recruit about 4,000 men, which President Taft, under public pressure, soon augmented with 4,000 troops, including seven companies of Buffalo Soldiers. By mid-August, another storm ignited thousands of more fires. On August 20, winds picked up strength, reached hurricane force, and pushed what was now close to 3,000 fires into one huge conflagration. It was impossible to contain.

The 1910 fire season emphasized the fact that humans needed to rethink their efforts. Despite Pinchot's strongly held belief that forest fires could be controlled, the 1910 fires proved how puny humans are in the face of such a blaze. All told, some 10,000 firefighters—the largest fire-fighting force ever assembled—were recruited at the last minute in response to exceedingly dry conditions and thousands of fires in Idaho and Montana. They were almost entirely unprepared for the job. Most had no experience in forests. They had inadequate boots and clothing. They often lacked even rudimentary tools such as axes and shovels. Congress had not provided sufficient funding for basic training and equipment, let alone housing and food. "After sending out a rescue call to save the national forests," writes Timothy Egan,

> the government enlisted college boys from California; day workers from Denver, Salt Lake City, Butte, Missoula, Spokane, Seattle, and Portland; and immigrants—thousands of foreigners—from mining camps in Arizona and Colorado, from irrigation ditches in California, from timber towns in the coastal Pacific Northwest, people working at jobs American citizens would not take.[15]

In fact, there was very little such a huge fire-fighting force could do in the face of so many fires, which finally blew up into the massive, entirely uncontrollable blaze called the Big Burn. It is worth repeating, because scale is hard to grasp. Three million acres burned in two days. That's 4,687 square miles, almost the size of Connecticut.

Ranger Ed Pulaski showed heroism in saving 40 men in the midst of that fire. His actions have achieved near-mythic status in the Forest Service. In the aftermath of the fire, Pulaski was unable to secure any compensation for his severe injuries. He could not even secure small sums for the maintenance of grave sites commemorating those who had died, and whose charred bodies had been gathered and carried out of the destroyed forest and placed in the ground. Nearly blinded in one eye, his lungs compromised by heat and smoke, and badly burned, Pulaski returned to work because he had no other way to support his family. But he also took it upon himself to continue to care for the graves of men who had died, because the federal government refused to do so. The Forest Service could not even allocate a few hundred dollars to provide markers for the grave site until 1921, when Congress finally appropriated 500 dollars for that purpose.

Congress, in its budgetary wisdom, eventually doubled Forest Service funds for roads and trails, recognizing that suppressing forest fires required a quick response. A quick response required two things, both of which were expensive. It required real time discovery of a fire, and an infrastructure which would allow the rapid mobilization of a fire-fighting crew. Both Graves and Greeley put the Forest Service on the path to building infrastructure appropriate for fire suppression—roads, trails, communication facilities—hoping to match Pinchot's belief with on-the-ground results.

The requirements of real time discovery of fire and adequate infrastructure were further promoted by the fifth Chief Forester. Mindful of the consequences of large fires, Ferdinand Silcox instituted in 1935 a "10 a.m. Policy," which meant any fire spotted must be put out by 10 a.m. on the following day, no matter how remote or how large. To achieve this goal, the Forest Service scrambled to build lookout towers, roads, and trails to ensure any fires were extinguished as quickly as feasible. Eventually, over 8,000 lookout sites were established, located in every state except Kansas.[16]

With real time discovery and appropriate infrastructure, the suppression of fire began in earnest, and continued in earnest. The Forest Service became the agency that fought fires. As fire historian Stephen J. Pyne remarks of the Forest Service:

> From its origins, fire control was not just something the agency did among its varied tasks. Perhaps more than anything else, fire was the reason the agency existed at all, and it became a primary index of the agency's success.

He goes on to note, "Its origin stories were mostly forged in flame." By the 1960s, origin stories, massive funding, and an impressive organization geared toward fighting fires, resulted in what Pyne calls "the premier wildland fire agency on Earth."[17] That premier agency fought fires and cut trees.

One contradiction bordering on the absurd was that within the agency, suppression of fire, however impressively effective, was accompanied by the suppression of knowledge about fire. Whether raised by scientists or the public, any discussion of the benefits of fire in forests, if at odds with the prevailing orthodoxy, were to be ignored or disputed.

Suppression of fire and suppression of knowledge about fire were not absolute, however. Intentional burning was allowed on southeastern forests in the 1930s and 1940s, supported by Forest Service and other research which indicated the importance of fire in longleaf and loblolly pine forests. But the Forest Service nonetheless kept both the science and the practice hidden as best it could, first by ignoring early research, such as H.H. Chapman's research on fire, published between 1909 and the early 1940s, and second by refusing to let intentional burning occur anywhere except in southern forests, with limited exceptions. Third, the Forest Service kept hidden its own research on what eventually became known as the ecology of fire.[18]

For a half century, despite some intentional but reluctant burning in southeastern forests, the Forest Service developed policy to exclude fire from national forests, and also to provide funding to states to similarly exclude fire on state forests, based on the 1924 Clarke-McNary Act. Forest Service policy to suppress and exclude fire from forests remains a rather extreme example of how an *imagined* order of the natural world organized the efforts of the bureaucracy. Those who thought the world was otherwise, and whose research or cultural experience pointed toward fire as important for the health of a landscape, were overlooked and marginalized.

For "zealous technocrats" within the Forest Service, writes historian Ashley L. Schiff, "purpose was transformed into a mission, a campaign into a crusade." Fire was the enemy. Just as Hays had couched the question of Progressive Era conservation in evangelical terms—a gospel of efficiency—so Schiff recognized something of the unquestionable, indeed doctrinal nature of Forest Service fire suppression. "Thus had evangelism subverted a scientific program, impaired professionalism, violated canons of bureaucratic responsibility, undermined the democratic faith, and threatened the piney woods with ultimate extinction."[19]

Bureaucratic, doctrinal suppression of fire meant more timber went to market, and timber professionals were intent on supplying that market. It wasn't until the 1960s that ecological considerations came to the fore, including the idea of the importance of fire as a natural process. Managers needed to understand and accommodate this new approach. Timber professionals needed to realize that fire was not simply a scourge to extinguish. Zealous technocrats needed to rethink their crusade.

The 1963 Leopold Report, which touched on fire ecology, was an important early contribution to understanding landscapes as ecosystems.[20] Even earlier was Aldo Leopold's 1949 *A Sand County Almanac*, which discussed the importance of fire on the land. Earlier still was his essay "'Piute Forestry' vs. Forest Fire Prevention," published in 1920.[21] But Leopold, with his holistic, transcendental approach, failed to influence the timber professionals of the day.[22]

While scientists themselves informed the process of timber management, the central authority decided which scientists to believe and which to ignore. Forestry at the time was equated with silviculture. The results were ecosystems simplified into systems of harvesting and growing trees. Throughout the first half of the twentieth century, most rangers working on national forests had received the same sort of training. They had also been thoroughly socialized through effective means of centralized oversight. They saw their job as bringing trees to market, regenerating forests with desirable species, suppressing fires, and mitigating damage from insects. Simplifying the world in this way, scientifically informed foresters invented techniques of inventory and control, and devised a language of standardized forestry techniques. They participated in the growth of an agency that rewarded compliance, promoted shared values, and developed a "culture of conformity."[23] Forest simplification became the model for managing natural resources, just as organizational simplification became the model for managing humans managing those resources.

"Fire Control Notes," a Forest Service publication begun in the 1930s (since transformed into a contemporary publication called "Fire Management Today"), provides a direct source of early beliefs about fire and fire management. Roy Headley, writing in the first issue, noted that "[o]ver a period of 30 years since the inception of organized effort to stop the fire waste of American natural resources, impressive advances have been made." He goes on to observe:

> Considerable body of knowledge of the arts and sciences involved has accumulated. Systems of organizing and managing human forces and mechanical aids have in some instances attained dramatic efficiency. Fire research has won the respect of owners and managers of wild land. The advancement to date in technique entitles fire control to a place among the amazing technologies which have grown up in recent decades.[24]

The practical topics covered in "Fire Control Notes" range from which binoculars to use to how to install a periscope atop a fire lookout tower. In one essay, Regional Forester C.J. Buck defended against the complaints of hunters, the construction of roads and trails in forests as aids to suppressing fire. The "central aim" of roads in national forests in Oregon and Washington, he insisted, was for "the protection from fire at the lowest cost, of the 26 million acres within the National Forest boundaries of the two States." The road system was mapped out "based primarily on the control of fire within time limits." This meant that "every area in the National Forest has its limit set for the allowable time from the 'spotting' of a fire to the arrival on the ground of suppression forces." He couched the need for roads in terms of averting catastrophe:

> Fire suppression involves the first line of defense, consisting of one or two fire chasers for immediate action, and the second line of defense made up of varying sized crews drawn from outside and depending on larger trucks to speed them to the vicinity of the blaze. There is no romance in this grim business. The first hour or two after discovery of a fire usually determines whether the flames will be confined to a negligible area or break out into a real forest annihilating blaze.[25]

In 1930–1931, 20,825 miles of roads were planned; by 1936, 75 percent had been completed. Trails were similarly important for quickly reaching fires. Eighty-five percent of the planned 35,284 miles of trails had been built by this same time. Buck is careful to point out that these roads and trails were built to preserve forests, not for recreationists nor for the convenience of administration. He asks, "Shall we have protected forests with roads, or unprotected forests without roads?" His answer, which in practical effect was carved through forests in Washington and Oregon, demanded efficient fire suppression: "The protection of the forest demands speed in putting out fires, and speed in these days of motor vehicles means roads."[26]

Speed also meant rationalizing the movements of men on fire lines. "Everybody deplores the slow speed at which fire line is usually constructed," wrote Kenneth P. McReynolds. Slow speed allowed "many fires to get away because of the inefficiency of converting available energy into a held line." A "line" is a trench dug into the earth, a means to stop the progress of a fire across a landscape. The question was how many men with what sort of handheld equipment were needed for optimal production of a line. McReynolds staged trials on the Rogue River National Forest with 192 men, and again with 154 men, on a mock fire in an attempt to gauge efficiency on a 25–50 degree slope with variable vegetation. Axes, hoes, and shovels were among the tools used. The unit of measure was the number of "links" per man-hour needed to construct a line. In the first test, the speed of construction was 31 links per man-hour; in the second, it was 60 links per man-hour (a link is 1/100 of a chain; a chain is usually 66 feet long). The result was roughly 20 feet of line constructed per man-hour in the first test, and 40 feet of line per man-hour constructed in the second test. The point of this and many other studies was to dig the maximum line with the minimum of effort in the shortest time with the fewest men.[27]

A modern, ecological critique of fire suppression tends to overlook two significant aspects of forests during the first half of the twentieth century. One is environmental, the other cultural. They are not neatly separable.[28] Fire suppression created a certain kind of forest, no less than timber harvesting created a certain kind of forest. Those forests, protected and exploited, became known quite directly and physically through the work of thousands of men who learned about the nature of fire as they actively tried to extinguish it. They learned through their own physical efforts about terrain and water and trees and fire. They learned about weather. They walked to put out fires. They rode horses to put out fires. If roads had been built into the forest, they drove trucks to those fires. They scanned the horizon for smoke from remote lookouts. Some of them learned to parachute out of airplanes to put out fires in landscapes that were otherwise too difficult to access quickly. Their efforts were seen as heroic. Protecting forests was indeed a calling. The zealotry may well have "subverted a scientific program, [and] impaired professionalism," but it also created a cadre of men with a profound relationship to forests. That cadre of men, and a comparatively tiny number of women, remain obscure to contemporary employees of the Forest Service. This practice is itself characteristic of a bureaucracy: forgetting its own history and the effects of that history.

Mostly remembered is forest protection for a singular purpose: to cut a steady, sustainable supply of timber for the market. Then, as now, "efficiency" was a virtue, a means to avoid waste, and a forest in flames was the epitome of wastefulness. Second Chief Forester Henry Graves, in the *Principles of Handling Woodlands*, laid out what would become standard practice. He warned against allowing the natural world to take its own course, and insisted forests be organized for human purpose. "In the struggle for existence between two species," he wrote, "the one which has the least market value often wins."[29] To avoid this unhappy outcome, Graves promoted the construction of roads, trails, and telephone lines to facilitate fire suppression. "The first measure necessary for the successful practice of forestry," he insisted, "is protection from forest fires."[30]

In his influential book, Graves specified the object of silviculture in five assertions. In his view, silviculture is to (1) "secure quick reproduction after the removal of timber," (2) "produce valuable species instead of those having little or no market value," (3) "ensure a full stock, in contrast to stands of lesser yield," (4) "produce trees of good form and quality," and (5) "accomplish the most rapid growth compatible with a full stand and good quality."[31]

Efficient means to achieve these quick, rapid, and valuable forestry ends were needed because the market required efficiency. Graves frequently used a conventional agricultural framework and referred to trees as crops subject like other commodities to market forces. Since the time Graves penned his book on handling woodlands as agricultural products, reducing the world to efficient, measurable pecuniary considerations—reducing trees to marketable crops—has become fairly standard economic practice. It comes at the cost of obscuring other significant cultural and natural elements, however, that are neither easily quantifiable nor obviously market oriented. For some economists, multiple use of public land, which requires managers to attend to diverse values and activities, is itself inefficient because of a "conflict" of values that leads to "departures" from the transcendent goal of economic efficiency. Economic efficiency is simply the ideal state, and efforts in timber management, as in other areas of the market, should try to approach this culturally significant ideal. The first three Chief Foresters—Pinchot, Graves, and Greeley—and the fifth—Silcox—all satisfied this ideal by insisting on protecting forests from fire for timber harvest, and then by growing industrial forests with the most marketable species. Forest Service adherence to market solutions based on efficiency criteria would resonate with later economists. It would also cause environmentalists, with a different set of transcendent values, to shudder with despair.[32]

Sins of the Past

The shuddering eventually took center stage and resulted in the 1976 National Forest Management Act (NFMA), spearheaded by Minnesota Senator Hubert H. Humphrey. This act limited clear-cutting, required each national forest to develop a comprehensive management plan (periodically revised), and prohibited timber harvests if reforestation could not be reasonably started within five years. The act required careful analysis to ensure long-term sustainability of timber harvests, the forests themselves, and the fish and wildlife that inhabited them. It also reiterated the importance of NEPA and insisted that the principles of the 1960 Multiple-Use Sustained-Yield Act be followed. NFMA was another act of inefficiency because it required thoughtfulness. The emerging science of ecology was a prime mover and scientists were key players. Citizens and courts had important roles to play. The act was not easily ignored.

The Forest Service, however, often ignored the act and continued to put out for bid sales of timber in ways contrary to NFMA. The destruction of northern spotted owl habitat in northwest forests is a central example. Timber production remained an overarching concern for the Forest Service, as it paid insufficient attention to broader ecological considerations. As citizens complained, scientists conducted studies, and courts intervened, the Forest Service increasingly was required to change its aggressive timber-centric practices. The federal district court judge who ruled against the Forest Service in 1991 noted a "deliberate and systematic refusal" at the highest levels of the Forest Service and the Fish and Wildlife Service to follow the dictates of NFMA. Indeed, high-level executives in both agencies were found to have failed to "comply with the laws protecting wildlife."[33] The Forest Service, the district court ruled, could not simply ignore its responsibilities under various statutes because of its decades-long timber-harvesting ethos. The agency was no longer the final arbiter of its own behavior. Nor could it reasonably ignore shifting public sentiment that insisted on better environmental stewardship. With this ruling, the age of the Forest Service as primarily a timber-management agency had come to an end.

My pitch for inefficiency, in response to the 2017 requirement to cut more timber on national forests, was also a mild caution to remember past sins. A timber-centric Forest Service was clearly a thing of the past. Simplified industrial forests as models for management had become complex natural forests once more, with ecosystem management the new rallying cry. Timber was one among many values we needed to shepherd in the administration of public lands. Trying to speed up—to make more efficient—all aspects of timber production cut against the grain of public interest. It may well result in court intervention if citizen involvement, required under NFMA and the Forest Service's own 2012 planning rule, is truncated, and scientific understanding is ignored. That planning rule provides, as the rule itself states, "a process for planning that is adaptive and science-based, engages the public, and is designed to be efficient, effective, and within the Agency's ability to implement."[34] But here as elsewhere, gospels of efficiency are hard to shake loose, especially if reanimated by political appointees pursuing economic agendas and deregulatory reforms. EADM—environmental analysis and decision making—the Trump administration's attempt to undermine NEPA by making it more "efficient," may well be the ghost of efficiency past, come back to haunt us.

The Efficient and the Good

Bureaucratic thought tends to mimic economic thought, forever seeking efficiencies and developing measures of value which then become sources of value. But there remains hidden in all those fire-suppressed forests a more intricate cultural and historical context needed to understand efficiency and its valorization. How did a notion from engineering, first clearly articulated about the mechanics of waterwheels, come to characterize so much of social life?

One mid-nineteenth century definition of efficiency is "the ratio of useful work done to energy expended." With technical origins in physics and engineering, the term and idea of efficiency entered many realms of human thought and activity. Mechanical efficiency, organizational efficiency, economic efficiency, industrial efficiency, educational efficiency, household efficiency, personal efficiency, leisure efficiency—all were increasingly scrutinized to eliminate waste, minimize the energy expended, and increase the useful work done. Technical experts became key players and offered advice on "improvements" in a variety of mechanical and social domains. With Progressive-era advocates like Pinchot and Graves, timber production and national forest management were no exceptions. Technical experts should be accorded primacy; all others should stand back and watch the experts at work—or at least take expert advice on how best to work.

In *The Mantra of Efficiency: From Waterwheel to Social Control*, historian Jennifer Karns Alexander notes that "[e]fficiency is celebrated … [it] is also a deeply troubling idea."[35] Clearly the celebratory side of efficiency has infused the Forest Service, the troubling side has not. Alexander describes efficiency's adoption into many domains, such as Henry Ford's production lines, Frederick Winslow Taylor's organizational methodology, and Alfred Marshall's theorizing in his 1890 *Principles of Economics*. Her effort is to show efficiency's many guises and uses and to ask how a certain objectification, and a certain morality, became attached to efficiency. She cites Samuel Haber, who writes in his book *Efficiency and Uplift: Scientific Management in the Progressive Era*, "*Efficient* and *good* came closer to meaning the same thing in these years than in any other period of American history."[36]

The efficient and the good—one a technical measure, the other a moral evaluation. Yet increasingly the distinction seemed lost, as efficiency became culturally prominent. "Efficiency soon described diets, appliances, child rearing and education, clothing design, exercise, saving money, curing neurasthenia, and a path to global dominance," as Alexander reports. She goes on to observe:

> Efficiency itself became a prominent feature of advertising—of motors, soaps, and breakfast foods—and appeared in daily newspapers, home economics journals, and mechanics' magazines. By 1915 the word *efficiency* was plastered everywhere—in headlines, advertisements, editorials, business manuals, and church bulletins.[37]

In the early to mid-twentieth century, if one needed help chewing food, efficiency experts offered advice. If one pined for freedom, efficiency was the path. If one wanted to build muscle or strengthen resolve, it was time to start with efficient self-discipline. If one hungered for love, one could first take a self-test to gauge one's personal efficiency. And take care to measure results. Without an appropriate metric, judging efficiency was only a guess.

One widely read efficiency theorist of the day provided the moral grounding for greater personal efficiency:

> Efficiency is the difference between wealth and poverty, fame and obscurity, power and weakness, health and disease, growth and death, hope and despair. Efficiency makes kings of us all. Only efficiency conquers fate … Efficiency leads us from a world of chance to a realm of choice, changing us from automatons to men. Efficiency provides our only freedom—that of shaping circumstances and hewing events to suit ourselves![38]

Whether people hewed events or trees, the measure of efficiency became a source of moral authority, a guide to behavior and belief, and, apparently, to good health, if not good governance. Measures of efficiency may be improved, and new techniques of efficiency may be invented, but because efficiency also took on a patina of morality it became problematic to argue against the idea of efficiency.

The efficient and the good, bound together, became one. Inefficiency and waste, also bound together, similarly became one. The symbolism is as clear as it is simple and has proven long-lasting.

A glance at the contemporary *International Journal of Forest Engineering* shows that in the world of forestry "efficiencies" are sought in all domains, from patterns of cutting logs to managerial performance, all marked by measures with which to judge "optimal" forms and practices. Such scholarly work often provides novel ways to think about old problems. There is, after all, much to be gained from efficient machinery and efficient operations, although it is not at all clear that resource managers read these sorts of studies.

What Forest Service managers do read are documents such as the 2012 National Forest System Land Management Planning Final Rule and Record of Decision. An examination of this Rule shows the rhetorical importance of efficiency for centralized planning. The words "efficient," "efficiently," and "inefficient" occur 58 times over 114 pages. For forest planning purposes, "efficiency" characterizes a diverse set of quite different administrative functions. Frameworks, processes, responses, designs, prioritizations, adaptations, resolutions, costs, and management, all are distinct, all are subject to efficiency. All functions are made to appear equivalent through recourse to efficiency. Reading such government documents (which are not unique to the Forest Service) shows that "efficiency" continues to be plastered everywhere, a century after it first achieved cultural prominence.

Avoiding waste seems to be ubiquitously implied; embracing the good, its symbolic counterpart, is thus unavoidable. Little wonder that with more than a century lauding efficiency in so many domains, my pitch for inefficiency was unworthy of consideration. My pitch was, in effect, immoral.

Yet efficiency, as it turns out, is an idea that is not solely about machinery or operations. It is an idea that organizes, as the above examples show, a great deal of conceptual, administrative, and pragmatic effort, such that to question efficiency's efficacy is to court trouble. The trouble starts, as I've suggested, with the symbolic oppositional pairing, efficient/virtuous, inefficient/wasteful. To argue against efficiency is to imply the horrors of wastefulness. To argue for inefficiency is to ignore the virtuous.[39]

But beyond the symbolic oddity, trouble piles up with issues of scale and incommensurable values. "Efficiency," Alexander says, "is an industrial value, having been developed to assess the performance of the great machines that powered industrialization: waterwheels, steam machines, internal combustion engines." She goes on to note that efficiency can be out of place in a modern, postindustrial world "which emphasizes the plurality of values and is girded underneath by the new global economy, built on networks rather than hierarchies and dependent upon flexible organization rather than centralized planning."[40]

One significant study of the plurality of values, and the resulting challenges to efficiency, is anthropologist Anna Lowenhaupt Tsing's aptly titled *Friction: An Ethnography of Global Connection*, a study nominally focused on the rainforests of Indonesia and their commercial exploitation under Suharto's corrupt regime. Tsing is interested in the rainforest and its dramatic transformation, which left much of it in ruins. She is interested in local communities, similarly often left in ruins with the transformation of encompassing forests. And she follows across the planet diverse connections to these forests—economic, political, and cultural—in all their clumsy awkwardness. She is curious where they lead, these connections, and shows that awkward relationships between environments, local peoples, governments, capitalists, and environmentalists are often fraught, and far from efficient.

Tsing shows that issues of scale and incommensurable cultural values are not easily resolved. Her questions have broad applicability. Do we focus on the microsociology of local events? How then do we understand the interests of transnational environmental groups in Indonesian rainforests? Do we start with transnational environmental ideologies that seem to transcend location? How then do we resolve those ideologies within a local vernacular and relative to a local environment? Are the global economic processes that move through and incorporate local economies more significant than the local cultures that absorb and refashion those processes? How best do we describe the efficiencies of centralized planning and hierarchical social forms against the friction of far-reaching networks of environmental activists?

Conceptual resolution matters—in the sense of the relative sharpness of an image, and in the sense of resolving conflicts. A global scale offers a coarse resolution, of both image sharpness and conflict resolution, and often remains but an interesting abstraction from the viewpoint of local communities. A local scale offers a finer resolution but is mixed in terms of conflict resolution and broader concerns. Where, Tsing asks, is it best to focus attention? Is friction, not efficiency, simply a component of the larger system? Humans and their social forms, after all, are not waterwheels, for which the ratio of useful work done to energy expended is a viable measure.

Two poles in the where-to-focus debate can be seen in the work of anthropologists Eric Wolf, *Europe and the People Without History*, and Marshall Sahlins, *How "Natives"*

Think: About Captain Cook, For Example. Wolf argues that we need to focus on the economic forces that move through and transform local societies. Sahlins argues that we need to focus on local cultural understandings and responses to those larger economic forces.[41] Tsing says let's do both. For an adequate understanding of coupled human and natural systems, I'm with Tsing. Efficiency and friction, capitalist expansion and local cultural responses, nature devoid of humans and nature as anachronism—all are at play, and all are apparent in forests cut and sold and regrown and burned in the warming days of our planet.

Carbon, Proforestation, and Decadence

For timber harvesting in Forest Service planning, scale appears to be sliding, from local forest, to regional clusters of forests, to an overview of all national forests. But there it stops. Global concerns, say of carbon sequestration on a planet-wide scale, appear to not enter the equation of how much timber to cut. Supply chains—where timber goes once it is processed into usable lumber or some other product—also fail to enter the equation of how much timber to cut. The Forest Service wants logs on trucks, but where those logs go, and what happens in the event of a timber sale, are left for others to worry about. Bureaucracies, with limited fields of vision, limit the scale of their own involvement.

Yet it matters if trees—"biomass" in the term of the day—are burned to generate electricity, thereby releasing carbon, or are used in construction, thereby continuing sequestration of carbon. It matters if lumber or wood pellets or whole logs are shipped to another country for further processing. It matters economically, and it matters ecologically. The entire commodity chain, from harvest to transportation, is based on energy from fossil fuels. Carbon dioxide emissions from handheld chainsaws, large-scale harvesting equipment, transport trucks, wood products production facilities, and ships are all part of the carbon equation.[42]

The carbon equation is connected to solar energy, which strikes the earth across all the wavelengths of light. The earth radiates some of this energy as heat back into space in infrared wavelengths (800 nanometers to 1 millimeter). As it turns out, carbon dioxide molecules in the atmosphere absorb infrared energy. More atmospheric carbon dioxide molecules trap more infrared energy; fewer atmospheric carbon dioxide molecules trap less infrared energy. The effect is similar to glass trapping heat in a greenhouse, hence the term greenhouse gas. Burning fossil fuels, in conjunction with the release of other greenhouse gases such as methane, has dramatically increased in the last century and a half, contributing to rising temperatures, which in turn has led to melting glaciers, shifts in weather patterns, increases in extinction rates of a diversity of species, and food security issues, among other effects. Ocean levels and ocean temperatures rise as well, affecting all of earth's inhabitants, humans among them.

Recognizing these threats, the European Union, moving toward a goal of 20 percent renewable energy, has declared biomass to be a renewable and carbon-neutral source of energy, as opposed to coal, which is neither. The result has been an increase in demand for wood pellets burned in place of coal to generate electricity. The U.K. is now the largest importer of wood pellets, and the U.S. timber industry has supplied the new demand, mostly from forests in the southeast. Current models, however, show that burning biomass as a replacement for coal is not immediately carbon neutral, for two reasons. First, wood and coal have about the same carbon intensity, that is the same primary energy (0.027 vs. 0.025 tC GJ^{-1}), but the combustion efficiency of wood is lower than coal. To put it the other way around, coal is more efficient than wood in both processing and combustion efficiency. "Consequently," as John D. Sterman and colleagues write, "the first impact of displacing coal with wood is an increase in atmospheric CO_2 relative to continued coal use, creating an initial carbon debt."[43] The immediate effect of the large-scale burning of wood for energy instead coal appears to be the short-term release of more carbon into the atmosphere. Why is this a problem?

The initial carbon debt from burning wood for energy can be repaid by the regrowth of harvested forests and the consequent absorption of atmospheric carbon. However—and this is the second reason to question the short-term substitution of wood for coal—the reabsorption of atmospheric carbon from burning wood has a time lag of between 44 and 104 years depending among other factors upon the regrowth of the harvested forest. The time lag is important because in the interim the earth may exceed the 1.5–2.0 degrees Celsius range of increase specified in the 2015 Paris Climate Change Agreement. The rapid regrowth of certain kinds of forests may help with carbon dioxide removal, and thus help mitigate the overall effects of greenhouse gases. However, it appears that replacing hardwood forests with, for example, rapidly growing loblolly pine plantation forests actually increases atmospheric CO_2 because tree plantations have lower equilibrium carbon density than natural forests. Natural, intact, and stable hardwood forests simply sequester more carbon than pine plantation forests. "In sum," Sterman and colleagues write,

> although bioenergy from wood can lower long-run CO_2 concentrations compared to fossil fuels, its first impact is an increase in CO_2, worsening global warming over the critical period through 2100 even if the wood offsets coal, the most carbon-intensive fossil fuel. Declaring that biofuels are carbon neutral as the EU and others have done, erroneously assumes forest regrowth quickly and fully offsets the emissions from biofuel production and combustion. The neutrality assumption is not valid because it ignores the transient, but decades to centuries long, increase in CO_2 caused by biofuels.[44]

Where does this leave the Forest Service with its charge to cut more trees? Should it produce a running tally of carbon lost and carbon captured by its forests? The Forest Service has a Forest Inventory Analysis program with a forest carbon estimation tool, but it is unclear if, during the Trump administration, that program was robust or moribund. Its website was last updated November, 2015. The 2015 Paris Climate Change Agreement, with 192 signatories, recognized the threat of global warming and called for balancing human-caused atmospheric carbon emissions with carbon removal. After the U.S. withdrew from that agreement, Forest Service progress on updating forest carbon estimates appears to have stopped, at least as indicated by its primary website. A moribund program may well be the unfortunate, politically motivated result.[45]

And yet, despite politics, the Forest Service continues to publish its scientists' work. For example, the "Assessment of the Influence of Disturbance, Management Activities, and Environmental Factors on Carbon Stocks of United States Forests," published in November, 2019, provides a region-by-region picture of carbon stocks between 1990 and 2011, excluding Hawaii. The U.S. Geological Survey study, "Baseline and Projected Future Carbon Storage and Carbon Fluxes in Ecosystems of Hawai'i," published in 2017, fills in the carbon picture for Hawaii. The question, however, is how carefully—if at all—these sorts of new scientific assessments enter forest-level planning, when that planning is driven by the political demand to cut more trees.[46]

Forests worldwide capture and store about 25 percent of all anthropogenic carbon emissions. A more fine-grained understanding of particular forests in particular geographies is beginning to emerge. Many types of disturbances kill trees, stop their sequestration of carbon, and result in the eventual release of that carbon. The most significant forest disturbances are timber harvest, fire, wind, insects, drought, and land conversion. Achieving a carbon balance, where carbon emissions are equivalent to carbon storage, will require understanding the relative importance of these sorts of disturbances.[47]

In the U.S., excluding Alaska and Hawaii, timber harvests account for 85 percent of carbon loss from forests each year. Standard post-harvest reforestation with seedlings or seeds may not be quick or extensive enough to absorb sufficient carbon to slow global warming. Afforestation—the practice of growing forests on non-forest lands such as former farms—may also fail to sequester sufficient carbon, for the same reason. Global warming is too urgent to rely solely on reforestation—the Forest Service's preferred strategy—or afforestation. Recent evidence points to the addition of a third strategy. William Mooman and colleagues call this strategy "proforestation," which they define as the practice of growing existing forests to their full, ecologically diverse potential.[48]

The neologism is awkward but serves to make a point. Not all forests should be managed to yield timber. Not all forests should be transformed into plantations. And there are good reasons to keep some forests as intact as possible: to maintain biodiversity, and to mitigate against climate change. Yet proforestation immediately runs into several practical and ideological problems. A very small proportion of U.S. forests is managed for ecological complexity, and an even smaller proportion is managed to be intact. Somewhere between six percent and seven percent of total U.S. forest area in the Lower 48 States remain intact. These are mostly found in national parks, private holdings, and wilderness areas on Forest Service lands. Typically, management in these forests has excluded fire suppression, allowing natural, often lightning-caused fires to burn.

Prior generations of foresters have referred to aging, intact forests as "decadent," by which they mean older trees within older forests. Having reached the end of their lives, aging trees were considered "useless." Downed by wind, rotted from within by fungus, drilled full of holes by birds seeking an insect meal, or simply overly "mature," decadent trees got in the way of commercial interests. By contrast, younger trees were "vigorous." A tree's ideal commercial maturation age could be calculated, and its harvest could be planned in 20-, 40-, or 60-year rotations, for fast growing species, or longer rotations for slower growing species. The aim was to transform forests into quickly growing, tractable, and commercially valuable systems. This vision was to make forests mimic large-scale agriculture. Measurement, manipulation, simplification, harvest, and control proved more important than evolutionary forces and complex species interactions. In other words, economic modes of thought superseded ecological realities. From the early days of the Forest Service, management bent toward rationalizing forests, making them "efficient."

I have often thought that if one were to implement a similarly strident economic approach, the term "decadent" might equally apply to older bureaucracies. Many bureaucracies may have reached the end of their usefulness, are overly mature, have values that have eroded, and need vigorous new growth. Perhaps, for such organizations, a rotation cycle would be a vast improvement.

To manage forests to their full, ecological potential would require a significant shift in policy, which on Forest Service lands is for multiple uses, including timber harvesting. Recognizing conflicts of multiple use—one source of Tsing's "friction"—Mooman and colleagues make several policy suggestions. First, they insist policymakers understand that not all forests are "equivalently beneficial for a range of ecosystem services." Intact forests, as opposed to production forests, have a different set of benefits, including promoting species diversity and acting as long-term carbon reservoirs.[49] These benefits should take center stage in our current climate crisis. Second, those who drive policy, with their economic concerns, need to consider that intact forests are a cost-effective means to sequester and retain carbon for the long term. And third, policymakers should understand the approach could be applied to many forests of different types.[50]

Proforestation would then augment the assessment of "how much to cut, when," with the assessment of "how much not to cut, ever," even if these twinned assessments never achieve balance.

The Rationalized Forest

Early foresters were interested in sustained timber yield, in part because they saw landscapes devastated by commercial logging. Contemporary foresters are, or should be, interested in sustained ecological function, and for the same general reason: they should understand ecological consequences of past and present management practices. Yet we may not be able to sustain both timber yield and ecological function.[51] Sustained timber yield requires simplifying forests into industrial plantations. Sustained ecological function requires complexifying forests into something resembling a natural state. Will we have plantations or forests? This is of course a sharp, simplifying dichotomy, but bureaucratic resolution may require such simplification, even as ecological and cultural realities remain messy and contingent. The central problem, as James C. Scott, Anna Tsing, and others point out, appears when a bureaucracy mistakes its simplification for an adequate image of the world before it. To put it bluntly, reality is never what a bureaucracy imagines it to be.

The bureaucracy, with visions of efficiency, prefers a rationalized forest. But of what sort? Scientific reforestation techniques with origins in German forestry, essentially to cut down a mixed-species forest and replant it with rows of a single species, proved highly commercially successful until such techniques failed catastrophically and resulted in forest death (*Waldsterben*). It took a century for that failure to manifest and make clear that single species of trees of the same age aligned in rows were susceptible to insects, various diseases, and wind, among other calamities.

Such calamities followed efficient, rationalized forests. They were in fact the foreseeable result of rationalized forests: to destroy a forest, transform it into an efficient single species tree farm. Rows of trees, and tree harvests, became central to what was, in effect, an agricultural system driven by efficiencies, from which fire was excluded.

Yet fire, too, in all its unpredictability, eventually proved amenable to models of efficiency. The trick was to measure all that can burn in a forest, and to have that measure stand as proxy for the forest itself. Rationalization of forest "fuels"—all the material on and near the ground prone to burn—simplifies a forest, just as a single species plantation simplifies a forest, even if the intention, in both instances, is to improve the adaptability of forests to fire. Stephen Pyne points to what he calls the "dark side" of the National Fuel Inventory System, which, starting in the 1970s, "intellectually reduced landscapes to caches of combustibles and subtly allowed fuel appraisal to replace fire ecology." Pyne goes on to note that reducing forests to combustibles "shifted attention from fire's restoration as an ecological rehabilitation to fire's use to assist suppression." Moreover, he says,

> The modeling of fuels suited a discipline such as forestry, which had long prided itself on its mensurational skills: fuels treatment looked like silviculture by other means. Fuels could be weighed, counted, quantified. Fuel reduction could serve as a numerical index of prescribed burning's success in ways that amorphous "naturalness" could not—a hugely attractive proposition to an agency overturning a long-held policy. The fuels-fire nexus tended to support the old fire-control agenda instead of a fire-restoration one. Besides, because so much of the fuels problem was the outcome of accelerated logging, better fire control was a means to support that harvesting.[52]

The current political model for sustained timber yield, or at least for cutting more trees, appears to be this: policy direction from on high yields its implementation on the ground. Simplification in the guise of "efficiency" articulates the entire process. The political demands, and the sorts of simplified, rationalized forests at stake, are clear: cut more trees and ignore the complexities of economic and ecological reality. Simply cut more trees. The implementation, much less clear, then becomes bureaucratic, not ecological, as the Forest Service tries to accommodate political demand. As obvious as the absolute need to suppress fire once was, efficiency became the mantra of the day. As with fire and the suppression of knowledge, so with efficiency and the suppression of knowledge. When science contradicts policy or is seen as getting in the way of implementing policy, the political response is to ignore or marginalize science.[53] When science requires thoughtfulness, and hence bureaucratic inefficiency, the response is to speed up science so it becomes a technique of the bureaucrats, not a quest for knowledge intended to inform management. In this way, friction is minimized.

In such contexts, scientific research and competing cultural values, when they emerge into open discourse, often result in conflict. Regional Forester Buck, addressing in 1935 what he referred to as "road haters," was responding to that conflict, with a plea for efficient means to suppress fires. His question, "Shall we have protected forests with roads, or unprotected forests without roads?" was moot, even as for the Forest Service "protection" meant only the exclusion of fire from forests, nothing more. A number of Forest Service scientists and others at that time disagreed with the absolute exclusion of fires from the landscape, but it took many decades for these alternative voices to be heard, and for the efforts of those *disagreeable* scientists to be recognized by the powers-that-be.

In our current tree-cutting mode, the question is "are we meeting our targets?" A "target" is a goal, and that goal, politically motivated, is 3.7 billion board feet for 2019, and even more as of 2020. I've heard forest supervisors say quite directly that the target is meaningless and impossible to meet, in any case, because of lack of staff and an absence of industry resources. They have also pointed out that cutting more trees requires building more roads, with the consequent environmental problems, from increased runoff to habitat fragmentation. I've heard regional foresters say in response, "I don't want to be told 'no.' Cut trees. Don't tell me you can't. Just cut trees."

All that tree cutting has an *internal* goal, to be recorded in databases and tallied up. Those tallies then serve as the marker of successful response to political direction. The on-the-ground results remain of local forest-by-forest concern, of course, but not of political concern. Still, if the nine regional foresters, each of whom is responsible for between two and 34 national forests, respond to political direction, then the chief forester, who is responsible for all 155 national forests, can show her responsiveness to such direction. The U.S. Department of Agriculture, which issued the direction to cut more trees under Secretary Perdue's influence, can then be placated by billions of board feet, not by the ecological health of forests nor by the climate health of our planet. Does our political system drive our bureaucratic system? It is a conjoined system of cascading effect, and the people within the bureaucracy, even if they know the results may be harmful, fulfill their "mission" and are rewarded thereby. The measure of value becomes the source of value.

Yet when politicians, forest managers, and scientists disagree, and when citizens fail to concur concerning values over resource exploitation, preservation, or some balance between, then the process of decision-making necessarily leaves the hands of politicians and central bureaucratic authorities and enters into a larger cultural swirl—an engaging, interesting, often frustrating mix of values, symbols, politics, industry, legality, landscapes, and science follows. One such manifestation is loosely referred to as the environmental movement of the 1960s. The movement was propelled in part by Rachel Carson's famous book, *Silent Spring*, published in 1962.[54]

The National Environmental Policy Act, which responded to the mix of values prompted by various environmental crises, asserts that central authorities must take into account views other than their own, because central authorities, such as the Forest Service, have made a mess of things. Citizens in particular must be heard.

On the other hand, and this may be a crucial if overlooked point, if the focus is on bureaucratic "efficiency" in an attempt to rationalize forests, the discursive swirl recedes, as those in control try to reassert their control. The question is not what to manage, real trees on real landscapes with real atmospheric carbon consequences. The question is how much efficiency to use. The issue of actual changes to landscapes and climates becomes subordinate to the bureaucratic processes to effect such changes. What's now at stake is the "the control (of the control) of nature," in Elizabeth Kolbert's phrase.[55] To control nature on a planet vastly transformed by humanity, Kolbert posits, will require ever greater control. If she is right, centralized knowledge and the means of its collection will become the bureaucratic focus, shifting from on-the-ground work to the interior work of the bureaucracy itself. This is where "efficiency" is king. This is the kingdom I called into question.

The Virtues of Inefficiency

I reminded the crowd of Forest Service leaders that for some environmental problems, indeed for some proposed sales of timber, there are no clear paths forward. Even the experts may be stymied. Scientific understanding may be incomplete, as is often the case in complex ecosystems. Policymakers, for their part, may be unable to define the issue clearly. Why cut 15 percent more trees? Indeed, policymakers may have no interest in defining the issue, simply commanding a preordained result. Stakeholders, from environmentalists to lumber mill owners, may articulate widely divergent opinions on what are called in Forest Service documents "desired conditions." Whose desires? What conditions? And resources managers, those with the real responsibility for managing real environments, may find their hands tied by political appointees following the unpredictable results of presidential elections. In such cases, complexity and uncertainty often render public decision-making especially contentious, whereby thoughtfulness and inefficiency then go hand-in-hand.[56] At the moment I became concerned that on the flip side, efficiency may well lead to an inadvisable rush to action. I was concerned that we were engaged in a headlong rush.

And so I continued to contend that "efficiency" limits choice by insisting on a subset of behavior and belief as germane to the task at hand, and then by exclusion insisting that other behaviors and beliefs do not belong. I suggested that we had become hell-bent on efficiently using "efficiency" in a way that made our larger goals irrelevant. While the Forest Service may have been all about cutting trees and growing industrial forests 50 and more years ago, our public lands and their uses were now no longer so easily simplified.

Human desires have changed. Our managed landscapes have changed. Our economies have changed. Our publics have changed. Our laws have changed. Our sciences have changed. Conditions have changed.

I briefly mentioned my fondness for fly fishing, which I found to be a lovely way to immerse myself in nature, but which must be the least efficient means to fish humans have invented. A weir and a spear would efficiently catch trout on the small streams I favored. My casting often did not. My point was about the potential beauty of inefficiency, not easily quantified. But it was also slowing down the rush to cut more trees. I thought about mentioning that speed reading was a craze when I was in high school, a means to make reading more efficient. But it had the effect of missing much of the content and nearly all the nuance of a written work. Speed timber sales are the ecological equivalent to speed reading.

One leader took the microphone. Efficiency was the only way we could possibly function, and my suggestion was the height of foolishness. We were charged with increasing timber production. Our very own processes should not get in the way. Inefficiency? Bah! A fetish? Humbug! This rising leader was deeply offended by my remarks. It became clear that he adamantly contested my claims. I had come up against the classic philosophical risk of ostracism, based on ideas that challenge conventional wisdom.

Being an anthropologist working in a land management bureaucracy over the last decade, I question its constructs from time to time. I hold at arm's length for inspection our theories of cause and effect and function, and then wonder out loud how, when, and why our curiosity fails, to be replaced by collective acquiescence. I always seek to combine social and natural considerations in all our decisions, avoiding the unhelpful view that there exists some sort of natural world outside of and independent from humans, much as I dismiss the opposing, equally unhelpful view that human beings are all that matter.[57]

Some members in the meeting simply tolerated my comments, having listened to my off-center approach during the past few years. Besides, they would soon return to their forests of concern, relatively isolated from regional and national influences. They would do whatever they could to keep cutting more timber. My proposal appeared truly nonsensical.

It may seem a simple matter to increase timber production on our nation's forests. It certainly seemed so for those who issued that edict. Our national forests have become overgrown, in large measure because of prior fire-suppression policies. Cutting timber and engaging in intentional burning—called "prescribed fire" in the lingo—are both rational responses to over-thick forests. But our forests, as it turns out, have spotty and limited markets, the result of all kinds of economic circumstances, and making public timber available for private sale does not always result in a sale. In some parts of our country, there are few buyers, and fewer mills. In many instances, national forests prepare timber sales, an involved process, and no one bids. Consequently, without first addressing economic barriers, a simple edict to cut more trees is not particularly useful. However much it drives a land management bureaucracy, "efficiency" does not address underlying economic issues. The result is not so much bureaucratic inefficiency as it is a simple lack of real-world awareness. Perhaps political asseveration works as propaganda—even if backed up by some sort of power to decide—but it is not so effective as a technique to cut real trees responsive to real markets.

I had asked at the same event whether the agency planned any before-and-after economic studies. Would an increase in 15 percent matter, and if so, in which states, on what forests, and to what local effect? Such information would be very helpful to better target where increased timber sales should occur, based on market considerations.

The unfortunate answer was no. There were no studies planned to gauge the actual economic effect of cutting more trees. There was only the edict to so cut, animated by presumption and presumptuousness, which is to say, animated by politics. Short-term gains seemed to preclude an analysis of long-term costs.

I didn't ask, out of my own ignorance, whether as a collective body of leaders we were aware of any options to consider as we endeavored to follow political direction. The idea of "natural climate solutions" was being floated at the time, which suggested several pathways to slow global warming by increasing carbon storage in forests. These solutions included improved forest management practices such as reduced-impact logging and extended harvest cycles, the latter to manage for optimum wood yield, not for optimum economic yield.[58]

And in support of biodiversity, someone might have asked whether we planned to engage researchers to analyze whether selective logging altered genetic variation of the selected tree species and, if so, what would be the consequences for those species. More broadly, and perhaps more importantly, we should have planned eDNA studies to assess the overall biodiversity of forests in pre- and post-harvest scenarios.[59] Only later did I learn to ask these sorts of questions. To date, I have not heard them addressed.

Since the Forest Service has had flat budgets over many years, it is quite a stretch to cut more trees, and some programs will have to be curtailed. Does this mean closing campgrounds? Eliminating hiking trails? Closing, because too expensive to replace, old and dangerous snowmobile bridges? Further privatizing public recreational amenities? Increasing fees? The answers are uncertain, played out on particular forests as supervisors try to meet the political demand for more logs on trucks. Local, timber-based economies may or may not improve. Citizen access to national forests may or may not be impacted. Fly fishers, with their inefficient ways, may or may not have clear streams in which to stand hip-deep as they contemplate a late-spring damselfly hatch.

My argument in support of inefficiency, I remember thinking, was that fateful tree falling alone in a forest. Only the birds and the squirrels and the tumbling waterfalls paid fleeting attention. At the end of the day, I recalled a line from Mary McCarthy's novel, *Birds of America*. "'Nature!' the *babbo* shouted. 'Nature! Don't be a goddamn fool! Nature is an anachronism.'"

What Are Trees?

In an early hearing before the Senate, the new Secretary of Agriculture in the Trump administration consistently referred to trees as crops. From this perspective, which equates timber production on national forests with agricultural business practices, efficiency is understandable. Businesses need to be efficient to improve profits. If the federal government is a business, then maximizing profits and, perhaps, improving customer service follow. With this in mind, I began following the Secretary's lead, calling trees "crops" at every opportunity, which had an interesting effect among some of my colleagues. They hated it. Many of them told me to cut it out. They reminded me that trees are much more and other than "crops," especially the ecologists, who cringed at the label. But "crops" was the operative word.

Well, if not crops, then what are trees? That question of course has wide scope. One recent and popular meditation on trees is a good place to start. I refer to Peter Wohlleben's *The Hidden Life of Trees*.[60] Wohlleben, a life-long forester, reconsidered his earlier practice of managing forests for optimal timber production. Initially he saw the value of a forest measured in board feet, and that his job was to ensure maximum productivity of commercial forests. In his mature work, he began instead to think about and manage forests based on ideas of the interconnections between trees, fungi, insects, birds, and vertebrates, and natural forces such as wind, rain, and sun. He had begun, in other words, to think ecologically rather than economically.

Trees, he writes, are social creatures. They do not stand alone, with a collection of individual trees making up a forest. Instead they are interdependent in myriad ways. Roots connect trees with other trees, either directly root to root or through the mycelium of fungi—the often massive underground organisms that spread out, sometimes for hundreds of acres, often growing into and thereby connecting the roots of many, many trees. The results are unexpected. Fungi, for example, allow trees to share nutrients. In a forest of connected trees, nutrients from one tree can nourish another tree through their fungal intermediaries. Fungi also allow trees to share information, for example certain sorts of insect attacks on one tree can be communicated to other trees, allowing them to take defensive measures. In exchange, fungi, which do not photosynthesize, derive their needed sugars from trees.

Trees benefit; fungi benefit. In the larger system, humans benefit as well. Through fungi, Douglas fir and birch trees in Pacific Northwest forests, for example, move carbon from tree to tree. When Douglas firs are shaded, they receive more carbon from birch; when in full sun, they receive less carbon from birch. Fungi themselves store carbon, making them, in concert with the trees, a major repository of carbon—their carbon transfer and storage increasingly central to combat global warming. Douglas fir, birch, and humans are thus interrelated through fungi.

In 1997, the journal *Nature* dubbed the phenomenon of interconnected trees and fungi the "wood-wide web," based on Suzanne Simard's pioneering research in the field. Since the interconnections are below ground, they are hard to investigate, made more difficult to study because of the tiny size of fungi hyphae, tubular structures which grow into and connect with tree roots at the cellular scale. The combination of fungi and root form mycorrhiza—from the Greek words for fungus (*mykês*) and root (*riza*)—a wonderfully complex structure of underworld plant communication and collaboration. Recent work has shown that the mycorrhiza relationships between trees and fungi are ancient, perhaps 450–500 million years old, and complex, involving a still unknown number of species of fungi and as many as 60,000 species of trees.[61]

New scientific techniques have provided a clearer picture of the relationships between fungi and trees. Of particular concern for the Forest Service is, or should be, ectomycorrhizal fungi—these are the predominant fungi with tree associations in North America. Although only two percent of all plant species have ectomycorrhizal relationships, 60 percent of all trees growing on the planet are associated with ectomycorrhizal fungi. Such relationships are important for two reasons, both of which help buffer the Earth's climate against human-caused changes. Ectomycorrhizal fungi assist trees to photosynthesize faster in response to higher atmospheric carbon dioxide concentrations and low soil nitrogen. Ectomycorrhizal fungi also suppress the respiration in the soil of decomposer organisms. The result of faster photosynthesis and decreased soil organism respiration is a reduction of atmospheric carbon dioxide concentrations.[62]

What effect would a 15 percent increase in timber production have on ectomycorrhizal fungi and its benefits for buffering climate change? We don't yet know. As with questions of economic benefit, questions of this sort of ecological benefit have gone largely unaddressed by the Forest Service.

In addition to using fungi in the wood-wide web, trees communicate in other ways. Wohlleben points to African umbrella thorn acacia trees. When giraffes feed on acacia leaves, the tree pumps toxins into its leaves to discourage the giraffes. At the same time, acacia trees release ethylene in the air, which signals to nearby acacias that giraffes are afoot; neighboring trees similarly begin to pump toxins into their leaves. For their part, the giraffes move beyond the effective range of air-borne communication, about 100 yards, before they start browsing again on acacia trees.

Trees also communicate with other species. Attracting insect pollinators is a well-known example, which benefits trees and insects. Elms and pines also communicate with a species of parasitic wasp, which they attract by releasing a species-specific compound when they are being eaten by leaf-eating caterpillars. These trees can distinguish the saliva of different species of insects that are eating their leaves, and then attract a particular species of parasitic wasp, which in turn lay their eggs inside the caterpillars. As wasp larvae grow, they consume and kill the caterpillars, which benefits the trees as well as the wasps (but not, obviously, the caterpillars).

Conifers release compounds known as phytoncides, which account for the strong scent of pine trees, but which also serve to "disinfect" the surrounding area, lowering concentrations of insect pests. Walnut trees are similar, which is why in gardens, people often place benches beneath walnut trees, where they can sit and enjoy an afternoon relatively free of mosquitos. Phytoncides also have a positive effect for human immune responses, and may account for the increasing popularity of "forest bathing," as city-dwelling humans find novel ways to interact with forests and reimagine the benefits of forests.[63]

In addition to his discussion of how trees communicate with each other and with other species, Wohlleben notes the effects of heavy rain on a forest—the sort of heavy rain and subsequent flooding that appears to be increasing in parts of North America the result of climate change. Forests that cannot absorb massive amounts of water can lose through erosion up to 2,900 tons of soil per square mile per year, washed away into streams and rivers. Normal soil-building processes in a forest produce up to 290 tons of soil per square mile per year. It is clearly not a sustainable difference, with soil erosion far outpacing soil production, especially in disturbed forests. By contrast, intact forests, with deep humus formed over centuries if not millennia, have the ability to absorb more water than disturbed forests, and only lose up to 14 tons of soil per square mile per year.

When we think about trees, in other words, we need to think about dirt. Farmers clearly think about soils. It behooves foresters to similarly understand the processes of soil formation and erosion that accompany forest management, as they seek to grow and harvest their "crops."[64]

Global Cooling, Global Warming, Humans, and Forests

When we think about trees and dirt, insects and fungi, ecosystems and global systems, we also need to think about humans—at least over the last half-millennium if not longer. To understand what's in front of us, we need to understand how we got here. The short term holds every forest manager's immediate attention as they follow political direction, but the *longue durée* provides necessary historical context to understand that hold and that attention. Politicians have famously short spans of attention. Forest managers cannot be so indulgent.

Trees sequester, or store, carbon, which in our world of industry and industrial forests has become a concern. "Concern" is a mild term to use in this context, since global warming is clearly linked to burning fossil fuels and an increase in atmospheric carbon dioxide. The reason for the increase in carbon dioxide is straightforward, even if the effects are complicated. Over the last 150 years or so, as humans increasingly burned fossil fuels—coal, oil, natural gas—for energy, the result has been an increase in the carbon dioxide in the atmosphere. A graph from the National Aeronautics and Space Administration shows this increase over time (Figure 1.2).

During the last 70 years, the dramatic increase in atmospheric carbon dioxide is correlated with a rise in global temperatures and consequent climate unpredictability.

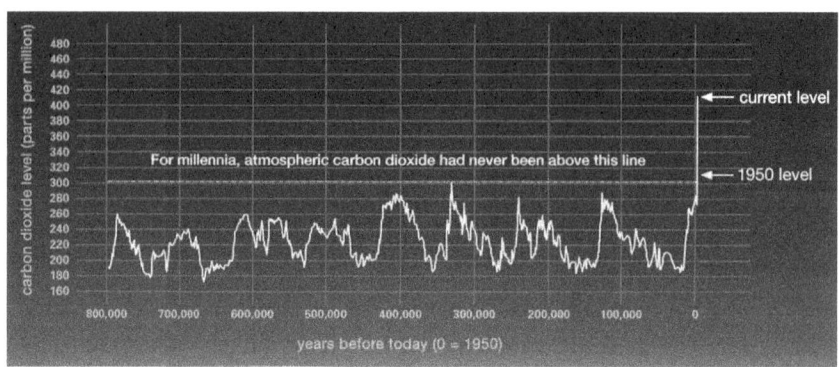

FIGURE 1.2 Atmospheric CO_2

Source: NASA.

Global-warming skeptics often point to the Little Ice Age as indicative of *natural* fluctuations of atmospheric carbon dioxide and a plunge in global temperatures by two degrees Celsius during the coldest period. The effects were most pronounced in Northern Europe. The causes of the Little Ice Age, however, which dates roughly between the early-fourteenth and mid-nineteenth centuries, have been hard to pinpoint. Sunspots and volcanoes may have played a role in global cooling and the decrease in atmospheric carbon dioxide during this time. Cooler oceans also absorb more atmospheric carbon dioxide than warmer oceans, compounding any cooling effect.[65]

The Little Ice Age may also have been partially—indeed, importantly—anthropogenic. European expansion, fire reduction, agricultural abandonment, the collapse of indigenous societies in the Americas, and reforestation may all be implicated. When Europeans arrived, the Americas were populated with somewhere between 45 and 78 million people, most of whom lived in the tropics and subtropics. Exact numbers are hard to come by. Recent estimates put the population in the Americas at over 60 million. (By comparison, China and Mongolia had an early sixteenth-century population of around 100 million, and Europe had a population of around 70–88 million.) European diseases, for which native peoples in the Americas had no immunity, killed 90 percent of them. Smallpox, measles, influenza, typhus and other diseases caused extraordinarily lethal epidemics. Such epidemics came in waves over more than a century and affected different parts of the Americas at different times. Some 54 million people died. Agricultural fields were abandoned, and forests regrew in their place.[66]

Antarctic ice core measurements indicate a drop of atmospheric carbon dioxide of about seven to ten parts per million after 1492—"an anomalously large decline," as one review notes. Isotope analyses indicate that concurrent with the decline in atmospheric carbon dioxide was a terrestrial uptake of carbon. The terrestrial uptake in carbon resulted, more likely than not, from reforestation of abandoned indigenous fields. The measurement of those fields and subsequent regrowth is tricky. The best estimate, taking into account seven regions of the Americas, is this: upwards of 55 million hectares of agricultural lands reverted to forests.[67]

The first point to emphasize is that humans influenced climate on a planet-wide scale prior to industrialization. As Alexander Koch and colleagues recently noted in their systematic review of the evidence, "The Great Dying of the Indigenous Peoples of the Americas resulted in a human-driven global impact on the Earth System in the two centuries prior to the Industrial Revolution."[68] Europeans, unaware of the mechanism of disease transmission, did not purposefully inflict diseases on indigenous peoples, even as they took advantage of the results. Unintended consequences are still consequences. Ninety percent of indigenous peoples died. Reforested agricultural lands, then, stored more carbon at a time of human population declines—and such declines were the result of pandemics accompanying European expansion into the Americas. That much is clear.

The second point is that humans have seen dramatic swings in atmospheric carbon dioxide over the last 500 years, roughly over 25 to 30 generations, which have corresponded with the Little Ice Age at one extreme, and our current global warming crisis at the other. The scale of forest regrowth and the consequences for carbon storage remain open to debate. But what climate-change skeptics don't acknowledge is the human-caused and massive shifts in carbon dioxide concentration in the atmosphere, clearly linked to global cooling and global warming, and other consequent weather changes. Arguing for "natural" causes to the Little Ice Age, such skeptics then extend their argument to our modern, warming planet as if it too results from "natural" causes. It turns out, they are wrong on both counts.

The third point is germane to current forest management practices, and to the political direction to cut more trees. The National Historic Preservation Act requires, among other things, that we take seriously our obligation to understand past landscapes and the humans who inhabited them. This is simultaneously a scientific and a moral obligation. The act in effect precludes us from blundering into forests, cutting as we go, without careful archaeological surveys and the resultant understanding of past lives—those whose societies and lifeways were entirely disrupted by European arrival. Without history in the long term, with only a focus on what is at hand, forest managers elide scientific and moral obligations in their endeavor to follow political direction. In so doing, they also run afoul of clear Congressional direction, that is, direction from the American people through their elected representatives to pay attention to the *longue durée*.

The Realm of Politics

None of this is to say that we should stop harvesting timber. Far from it. Cities built with wood may be our future, as they were once our past, and developments in cross-lamination, which produces structurally sound wooden beams and panels of great strength, are proving a match for steel and concrete, and at much lower energy and carbon costs to manufacture. Oak trees were once used to provide long and thick structural members for all kinds of buildings, from chapels to bridges. Pines, which used to appear limitless in their stretch across eastern North America, provided masts for ocean-going ships. Such forests are now mostly gone, cut over two and three and four times and scheduled to be cut again long before they reach their full potential as forests. Smaller trees are now engineered into forms that can rival steel and concrete as building materials, and new industries are emerging to take advantage of those new materials. But the science on how much to cut and when is still being sorted out.[69]

Forests, I've come to realize, are not only composed of trees on landscapes, in all their ecological complexity. Forests also exist in human structures, and if we can learn to think of our cities ecologically in the same way we are beginning to think of our forests ecologically, then we will be well on our way to ensuring the stability both. It is, of course, a stretch to envision such stability. But it is a vision worth pursuing, nonetheless. The Forest Service, with the right outlook, can help. It can help because it currently manages 193 million acres of public lands, much of which is forested. The decisions made by foresters matter. The point is to ensure that benefits to humans, and benefits to all the creatures that inhabit forests, are considered together in forest management.[70]

These considerations are not within the realm of science, or at least not solely in the realm of science. They are also, and importantly, in the realm of politics. Values are articulated in political discussions: values of our planet, human worth, economic standing, who exploits what and whom, and future possibilities. Who talks? Who listens? For some reason, our modern world has left it up to a teen-ager to become the voice of reason. Greta Thunberg of Sweden makes a practice of chastising the United Nations and world governments to take action on climate change. How powerfully mythic this situation appears. And how pathetic for those in control of world politics and world economies to get their comeuppance from a teenager and the world-wide demonstrations her efforts have initiated. As Thunberg has said many times, the science on climate change has been "crystal clear" for 30 years.[71] To adapt the Chinese proverb about planting trees, the best time to have acted on climate change was 30 years ago, and the second best time is now.

To make the promise of sustainable timber production a reality, humans must become less like a parasitic species, less like invading bark beetles, which destroy entire forests entirely for their own purposes, and more like a symbiont—more like ectomycorrhizal fungi, which cooperate with trees in the complex exchange of energy and nutrients. Humans must become part of the health of forests and the living world. This requires cooperation in many domains and necessitates careful scientific analyses no less than an alignment of economic interests, but not at the expense of the landscape itself (as Pinchot insisted). But it also requires something else, something capitalist economies lack, and anti-science politicians abjure, and which I describe as a practice of intellectual modesty.

By "modesty," I mean a recognition of the limitations of human knowledge, and a recognition of the coarse power of political ideologues who insist that their way is the only way. A sense of intellectual modesty requires listening to many, if not all, points of view about public lands and how to manage and preserve them. It requires openly acknowledging the advice of scientific experts, even if following that advice is not always possible. It requires explaining why an action was taken, why and how a grove of trees was cut, why an intentional fire was set, how one species was accorded primacy over another, or why a section of land was leased for mineral extraction or cattle grazing. It also requires understanding the kinds of power political appointees wield and pushing back against that power when warranted. Not anything goes. Even if the current Secretary of the Interior says so. Not all national forests can cut 15 percent more trees, even if the current Secretary of Agriculture makes that demand. Climate change needs collective action, even if intentionally misinformed senators insist otherwise.[72]

By "modesty," I don't mean a refusal to act, nor a refusal to question political direction, especially irrational political direction. As I see it, at an institutional level, the rational response to irrational policy is resistance. The irrational response to irrational policy is to try to implement it. But because peoples' livelihoods are intermingled with irrational policy, then from the perspective of the individual the reasonable response to irrational policy is to try to accommodate it—and variously so. After all, part of the purpose of promoting policies is to preserve one's job. In keeping with that purpose, many federal government employees in the U.S. are cheerfully apolitical, in the sense that every four or eight years a new administration will make changes to various agencies and their goals, and employees will go along adaptively. They see the arc of their careers in part as implementing whatever post-election oddments a new administration puts in place.

Sociologist Max Weber thought that bureaucrats attempt to make their agencies indispensable, not to the broader public, but to those who wield power. Hence, we find the presumptive apolitical stance of high-level bureaucrats. They kowtow to whomever is elected—including those elected under what may be considered odd circumstances, George W. Bush, for example, or Donald Trump. Even if contrary to the public good, or damaging of the natural world, new policies must be carried out. "Build that wall!" is one extreme example of policy stupidity, as if a wall between Mexico and the U.S. would do more good than harm. Clearly, ecological considerations were never part of the ethos of chanting crowds or tweeting presidents.[73]

Other examples are less clearly irrational. With increased timber cutting, employees in my agency variously agree to participate in the prevailing policy direction and the philosophy of public lands underlying it, as if it were reality. We pretend to agree, and in so doing make things happen. Yet policymakers manage to influence the on-the-ground outcomes of policy only insofar as others are recruited to help, and only insofar as the goals of the policy can be translated into others' goals and intentions. Policy by itself is ineffective no matter how often it is tweeted. Policy, to be effective, needs to be implemented. In the Forest Service, implementation happens at the forest level, and, within forests, at the district level. And even at this level of responsibility, agency employees are forced to make their efforts indispensable to those who wield power, regardless of environmental or social consequences, simply because of the personal consequences of resisting.

But if intellectual modesty is too much to ask of politicians and political appointees in positions of power, then perhaps we need something closer to humility. After all, managers of public lands—some 193 million acres for the Forest Service—manage those lands not for the bureaucracy and not primarily for this or that administration. They manage public lands for public benefit. One would think that hardly needs mention. But when the functioning of the bureaucracy becomes of paramount importance, and when this or that administration repurposes a bureaucracy for political ends, it does need to be said often, and insistently. Still, the clear peril for those within the Forest Service is to go against direction from political appointees whose "orders" travel through high-level civil servants down the "chain of command" to national forests and then to districts on those forests. Even the military language points to the bureaucracy, and not to landscapes, nor to the public whose landscapes we manage. I continue to insist, however, that landscapes and their public owners are the main issues, and I argue that land management bureaucracies often mischaracterize landscapes and their public owners in an attempt to transform both into something resembling a bureaucratic vision of rationality. And this is why I have the temerity—or maybe as my colleagues warn the poor judgement—to question efficiency, to object to speeding up scientific inquiry to support political direction, and to decry limiting citizen involvement for the same purpose. We need to be thoughtful. We need citizens to provide their thoughts about their lands. And we need to listen.

If modesty and humility fall short as management values, then perhaps an absence of hubris is the next best thing. The early chief foresters were all convinced they were right, that extreme suppression of fire was in the best interest of forests, timber economies, and people. And yet they were all wrong.

They managed nonetheless to build a huge fire-fighting apparatus. They built tens of thousands of miles of roads and trails to facilitate rapid response to fires. They built thousands of lookout towers to provide real-time warning of fires. They trained tens of thousands of men to fight fires. They eventually sent smokejumpers out of airplanes to stamp out fires. They educated the public to believe in fire suppression with successful advertising campaigns. And as a result of all this extraordinary effort, they grew forests that were prone to catastrophic fires (Figure 1.3). Their rationality is our legacy. Their hubris is ours to overcome.

"Trees," writes Colin Tudge, "are at the heart of all the necessary debates: ecological, social, economic, political, moral, religious."[74] Just so. But in 2020, we were in the grip of an anti-science administration that had dismantled, to greater or lesser extents, the very bureaucracies dependent upon science for adequate functioning. The Environmental Protection Agency, the Bureau of Land Management, and the Forest Service have all been undermined by the administrators charged with fulfilling their missions—Scott Pruitt, Ryan Zinke, and Sonny Perdue.[75] With these administrators in charge (and in Pruitt's and Zinke's cases, after their ousters for corruption, with their successors in charge), ecological debates shifted away from

FIGURE 1.3 "Your Forests—Your Fault—Your Loss!"

Source: U.S. Forest Service.

science to other realms, and the informational underpinning of this or that decision became lost to ideological concerns. Such ideological concerns are usually business concerns, not concerns over landscapes or people, and are as far from exhibiting any humility as one could imagine. Former EPA administrator Christine Todd Whitman, surveying the bureaucratic scene under the Trump administration, had this to say: "Right now, any [scientific] finding that seems to be restricting business, especially the energy industry, appears to be destined for elimination."[76]

The moral debate in particular needs to be reengaged around forests, as our planet warms and we imperil our future. Yet a morality—or better, a plurality of moralities—informing our collective future seems to be lacking in much of our political discourse. Still, we cannot follow former Montana Senator William A. Clark in his belief that, "Those who succeed us can well take care of themselves." As we ensure those who succeed us have a planet worth inhabiting, trees may be at the heart of all the necessary debates.

The Story of Forests

When I started working for the Forest Service, everyone seemed to be asking the same question. How do we tell the story of national forests? That question was posed by an impressive range of people, from the chief forester, to all the deputy chiefs, to the national directors of specific programs, to the regional foresters, to the regional directors, to forest supervisors, to engineers and recreation technicians on districts, and almost everyone else. The question still circulates. I was puzzled when I first heard that question, and I remain puzzled by the fact that it remains unanswered. Perhaps the question unites everyone in the agency. Perhaps that is why it is never answered. It functions not as a question, but as an assertion of uncertain identity.

To me, the answer is quite clear. The single most important story is this. The Forest Service, with good intentions, built a system to exclude fire from national forests. States and private landowners followed suit. The Forest Service also built a system to harvest trees from national forests, with uneven results over the last century. As a consequence of past management mistakes, the forests we have inherited are overgrown and thick with flammable understories. We now know fire exclusion is a problem.[77] We now know plantation-style industrial forests do not mimic natural forests. We now know we need to preserve ecological function first, timber production second, and avoid large clear-cuts. We now know fire-adapted landscapes are ecologically important and go a long way to ensure communities near forests are not consumed in the event of wildfire. We now know the importance of forests in carbon sequestration and in forestalling even more rapid climate change. We now know we need to shift away from concrete and steel to timber-based construction, and to judiciously burn wood instead of fossil fuels for energy.

Our modern management efforts are to thin—not cut down—forests and to encourage—not preclude—fire-adapted landscapes. Our timber production, and our intentionally set fires, is directed at long-term solutions to the problems caused by our ancestors. We see our efforts as experiments in public lands management, which will require constant adjustments as we continue to learn from the results of those experiments. Along the way, we welcome everyone's help because we don't have all the answers. But we also know, through past and present experience of public lands management, that this work is conceptually difficult, politically fraught, environmentally consequential, physically demanding, and labor intensive. And, as it turns out, none of it is efficient.

Notes

1 Gifford Pinchot, *Breaking New Ground* (New York, NY: Harcourt, Brace and Co., 1947), 506.

2 Samuel P. Hays, *Conservation and the Gospel of Efficiency: The Progressive Conservation Movement, 1890-1920* (Cambridge: Harvard University Press, 1959), 39–42. See also Herbert Kauffman, *The Forest Ranger: A Study in Administrative Behavior* (Baltimore, MD: Johns Hopkins University Press, 1960). James Q. Wilson's *Bureaucracy: What Government Agencies Do and Why They Do It* (New York, NY: Basic Books, 1989) rounds out the trinity of classics. On Muir and debates over natural resources see R. Nash, *Wilderness and the American Mind* (New Haven, CT: Yale University Press, 1979), Michael L. Smith, *Pacific Visions: California Scientists and the Environment, 1850-1915* (New Haven, CT: Yale University Press, 1987), and S. Fox, *The American Conservation Movement: John Muir and His Legacy* (Madison, WI: University of Wisconsin Press, 1981).

3 John Muir, "The Wild Parks and Forest Reservations of the West" *The Atlantic Monthly* (January 1898).

4 For the letter, see "The Principal Laws Relating to the Establishment and Administration of the National Forests and Other Forest Service Activities," *USDA Forest Service* (1964), 67.

5 Richard White, *Railroaded: The Transcontinentals and the Making of Modern America* (New York, NY: W.W. Norton & Company, 2011), 24. Today, fewer than 100 families own around 42 million acres. See J. Turkewitz, "Who Gets to Own the West?" *The New York Times* (June 22, 2019).

6 On Pinchot, Muir and Roosevelt, see Timothy Egan, *The Big Burn: Teddy Roosevelt and the Fire that Saved America* (New York, NY: Houghton Mifflin Harcourt Publishing, 2010). The Clark quote is on 48–49.

7 Hays, *Conservation and the Gospel of Efficiency*, 271.

8 Henry Solon Graves, *The Principles of Handling Woodlands* (New York, NY: John Wiley and Sons; London: Chapman and Hall, Limited, 1911), 3.

9 Graves, *The Principles of Handling Woodlands*, 6.

10 For a contrary view, see George T. Morgan, Jr., *William B. Greeley: A Practical Forester* (Minneapolis, MN: Lund Press, 1961).

11 William B. Greeley, *Forests and Men* (New York, NY: Doubleday & Company, 1952), 118.

12 See for example, William B. Greeley, "'Piute Forestry' and the Fallacy of Light Burning," *The Timberman* (March 1920).

13 Greeley, *Forests and Men*, 23.

14 Greeley, *Forests and Men*, 24, 26, 29.

15 Egan, *The Big Burn*, 8–9.

16 On the construction of lookout towers, see John R. Grosvenor, "A History of the Architecture of the USDA Forest Service," *USDA Forest Service* (July 1999).

17 Stephen J. Pyne, *Between Two Fires: A Fire History of Contemporary America* (Tucson, AZ: University of Arizona Press, 2015), 4, 7.

18 See, for example, H.H. Chapman, "Factors Determining Natural Reproduction of the Longleaf Pine on Cut-Over Lands in La Salle Parish, La." *Yale University School of Forestry*

Bulletin 16 (1926); "Is the Longleaf Type a Climax?" *Ecology* XIII:4 (October 1932). Anthony Godfrey's worked focused on California is also useful, *The Search for Forest Facts: A History of the Pacific Southwest Forest and Range Experiment Station, 1926-2000* (Albany, CA: USDA General Technical Report PSW-GTR-233, 2013).

19 Ashley L. Schiff, *Fire and Water: Scientific Heresy in the Forest Service* (Cambridge: Harvard University Press, 1962), 115. See also Randall O'Toole, "Money to Burn: Wildfire and Budget," in *Wildfire: A Century of Failed Forest Policy*, George Wuerthner, ed. (Washington, DC: Island Press, 2006), 217–221.

20 "Wildlife Management in the National Parks: The Leopold Report." www.nps.gov/park history/online_books/leopold/leopold.htm.

21 Aldo Leopold, *A Sand County Almanac: And Sketches Here and There* (Oxford: Oxford University Press, 1949), "'Piute Forestry' vs. Forest Fire Prevention," *Southwestern Magazine* (March 1920).

22 The Forest Service eventually acknowledged that it failed to listen to various scientists about the importance of fire in ecosystem processes. In the "Interagency Strategy for the Implementation of Federal Wildland Fire Management Policy" (June 20, 2003), for example, we find this:

> Fire exclusion efforts were intensified through the 1950s ... Some, however, were observing the adverse effects of fire exclusions on ecology and forest health as early as the 1920s. Men like Aldo Leopold in the Southwest, Harold Weaver in central Oregon, and H. L. Stoddard in the South were pioneers in questioning a fire policy based on fire exclusion. Despite their warnings, however, federal policy remained centered on putting out fires for the next several decades.

23 Kauffman, *The Forest Ranger*, remains the central study of the first half century of the Forest Service. His descriptions of reward, punishment, and socialization, bear careful reading. Graves, *The Principles of Handling Woodlands*, notes these aspects of silviculture, p. 6 ff. For a comparative perspective, see K. Sivaramakrishnan, *Modern Forests: Statemaking and Environmental Change in Colonial Eastern India* (Stanford, CA: Stanford University Press, 1999). "Culture of Conformity" is Stephen J. Pyne's phrase, echoing Kauffman; *Between Two Fires*, 17.

24 Roy Headley, "Fire Control Notes Offers Its Services," *Fire Control Notes: A Publication Devoted to the Technique of Forest Fire Control* 1(1) (December 1936), 3.

25 C.J. Buck, "Forest Roads or Forest Fires," *Fire Control Notes: A Publication Devoted to the Technique of Forest Fire Control* 1(1): (December 1936), 31, 32.

26 Ibid., 33. I know of no environmental history that traces the ecological consequence of all this early twentieth-century road building. Timber harvests and fire have had most attention, as in Nancy Langston's *Forest Dreams, Forest Nightmares: The Paradox of Old Growth in the Inland West* (Seattle, WA: University of Washington Press), and Steven Pyne's *Between Two Fires*.

27 Kenneth P. McReynolds, "Speeding Up Fire-Line Construction by the One-Lick Method," *Fire Control Notes: A Publication Devoted to the Technique of Forest Fire Control* (December 1936).

28 Richard White explores the intersection of work, environment, and environmentalism in "Are You an Environmentalist or Do You Work for a Living?: Work and Nature," in *Uncommon Ground: Toward Reinventing Nature*, William Cronon, ed. (New York, NY: W. W. Norton and Company, 1995).

29 Graves, *The Handling of Woodlands*, 9.

30 Ibid., 225.

31 Ibid., 8. Later in his book, he added esthetic value, silviculture value, and logging conditions as also important to address; 41–42.

32 See for example William F. Hyde, *Timber Supply, Land Allocation, and Economic Efficiency"* (Baltimore, MD: The Johns Hopkins University Press, for Resources for the Future, Inc., 1980). Hyde, perhaps representative of many economists, is steadfast in his belief about markets and market efficiency. For Hyde, the goal of economic efficiency is a given; departures from it a problem. From this perspective, multiple use is one such problem. Randal O'Toole, also an economist, provides a more nuanced and critical perspective, *Reforming the Forest Service* (Washington, DC: Island Press, 1988). Jim Furnish, architect during the Clinton administration of the roadless area rule protecting some 58 million acres of forest, traces his evolution from advocate for fire suppression and timber harvest to advocate for a more ecologically sound approach to forest management. *Toward a Natural Forest: The Forest Service in Transition* (Corvallis, OR: Oregon State University Press, 2015).

33 W.L. Dwyer, 1991. Order on Motions for Summary Judgement and for Dismissal. Seattle Audubon Society et al. v. F. Dale Robertson et al. No. 89099. U.S. District Court, Western District of Washington.

34 For the planning rule, see Federal Register Vol. 77. No. 68 (April 9, 2012).

35 Jennifer Karns Alexander, *The Mantra of Efficiency: From Waterwheel to Social Control* (Baltimore, MD: The Johns Hopkins University Press, 2008), 1.

36 Samuel Haber, *Efficiency and Uplift: Scientific Management in the Progressive Era* (Chicago, IL: Chicago University Press, 1964), ix. Emphasis added.

37 Alexander, *The Mantra of Efficiency*, 77.

38 Quoted in Alexander, *The Mantra of Efficiency*, 95.

39 Symbolic dualism is a venerable topic in anthropology. See Rodney Needham, ed., *Right and Left: Essays on Dual Symbolic Classification* (Chicago, IL: University of Chicago Press, 1973). See also Chris McManus, *Right Hand, Left Hand: The Origins of Asymmetry in Brains, Bodies, Atoms, and Cultures* (Cambridge: Harvard University Press, 2002).

40 Alexander, *The Mantra of Efficiency*, 148.

41 Eric R. Wolf, *Europe and the People Without History* (Berkeley, CA: University of California Press, 1982), and Marshal Sahlins, *How "Natives" Think: About Captain Cook, For Example* (Chicago, IL: University of Chicago Press, 1995).

42 See D. Markewitz, "Fossil Fuel Carbon Emissions from Silviculture: Impacts on Net Carbon Sequestration in Forests," *Forest Ecology and Management* 236:153–161 (2006), and E. O'Neill, R.D. Bergman, and M.E. Puettmann, "CORRIM: Forest Products Life-Cycle Analysis Update Overview," *Forest Products Journal* 67:308–311 (2017).

43 John D. Sterman, Lori Siegel, and Juliette N. Nooney-Varga, "Does Replacing Coal with Wood Lower CO_2 Emissions? Dynamic Lifecycle Analysis of Wood Bioenergy," *Environmental Research Letters* 13:8 (2018)[this is the page number of quote]. Economic analyses indicate that replacing coal with wood is not economically viable, without heavy government subsidy. See B. Mei and M. Wetzstein, "Burning Wood Pellets for US Electricity Generation? A Regime Switching Analysis," *Energy Economics* 65:434–441 (2017).

44 Sterman and Nooney-Varga, "Does Replacing Coal with Wood Lower CO_2 Emissions? Dynamic Lifecycle Analysis of Wood Bioenergy," 8.

45 www.fia.fs.fed.us/forestcarbon/index.php. Accessed 01/09/2020.

46 Richard A. Birdsey, Alexa J. Dugan, Sean P. Healey, Karen Dante-Wood, Fangmin Zhang, Gang Mo, Jing M. Chen, Alexander J. Hernandez, Crystal L. Raymond, James McCarter, *Assessment of the Influence of Disturbance, Management Activities, and Environmental Factors on*

Carbon Stocks of U.S. National Forests. Gen. Tech. Rep. RMRS-GTR-402 (Fort Collins, CO: U.S. Department of Agriculture, Forest Service, Rocky Mountain Research Station, 2019), 116 pages plus appendices. P.C. Selmants, C.P. Giardina, J.D. Jacobi, and Z. Zhu, eds., *Baseline and Projected Future Carbon Storage and Carbon Fluxes in Ecosystems of Hawai'i: U.S. Geological Survey Professional Paper 1834* (2017), 134 p., https://doi.org/10.3133/pp1834.

47 N.L. Harris, S.C. Hagen, S.S. Saatchi, T.R.H. Pearson, C.W. Woodall, G.M. Domke, et al., "Attribution of Net Carbon Change by Disturbance Type Across Forest Lands of the Conterminous United States," *Carbon Balance and Management* 11:24 (2016).

48 William R. Mooman, Susan A. Masino, and Edward K. Faison, "Intact Forests in the United States: Proforestation Mitigates Climate Change and Serves the Greatest Good," *Frontiers in Forests and Global Change* (June 2019); see also Jason M. Funk, N. Aguilar-Amuchastegui, W. Baldwin-Cantello, J. Busch, E. Chuvasov, T. Evans, B. Griffin, N. Harris, M. N. Ferreira, K. Petersen, O. Phillips, Muri G. Soares, and Richard J.A. van der Hoff, "Securing the Climate Benefits of Stable Forests," *Climate Policy* (April 2019). The recent National Academies of Sciences, Engineering, and Medicine report is useful, *Negative Emissions Technologies and Reliable Sequestration: A Research Agenda* (Washington, DC: The National Academies Press, 2019).

49 Some researchers insist on a change of terminology, reserving the term "forest" for multispecies "natural forests," and the term "plantation" for species-poor, if not monocultural, plantations that have replaced natural forests. See Anand M. Osuri, A. Gopal, T.R. Shankar Raman, R. DeFries, Susan C. Cook-Patton, and S. Naeem, "Greater Stability of Carbon Capture in Species-Rich Natural Forests Compared to Species-Poor Plantations," *Environmental Research Letters* 15:3 (2020).

50 Mooman, et al., "Intact Forests in the United States: Proforestation Mitigates Climate Change and Serves the Greatest Good."

51 Despite many attempts to model "ecosystem services" and to juggle priorities among them. The fatal flaw in this sort of reasoning is to compare incommensurables, which in logical style is similar to economic thought founded on cost/benefit analysis, for which all "values" are somehow monetizable. See, e.g., W. Scott Schwenk, Therese M. Donovan, William S. Keeton, and Jared S. Nunery, "Carbon Storage, Timber Production, and Biodiversity: Comparing Ecosystem Services with Multi-Criteria Decision Analysis," *Ecological Applications* 22(5) (July 2012).

52 Pyne, *Between Two Fires*, 127–128.

53 On how various administrators in the Trump administration have marginalized scientists, see Oliver Milman, "The Silenced: Meet the Climate Whistleblowers Muzzled by Trump," *The Guardian* (September 17, 2019), Adam Federman, "How Science Got Trampled in the Rush to Drill in the Arctic," *Politico* (July 26, 2019), Brittany Patterson, "Government Scientist Blocked from Talking About Climate and Wildfires: Critics Are Accusing the Trump Administration of Stifling the Dissemination of Taxpayer-Funded Science," *Climatewire* (October 31, 2017), Dina Fine Maron, "Trump Administration Restricts News from Federal Scientists at USDA, EPA: The Curbs Echo What Happened in Canada Six Years Ago," *Scientific American* (January 24, 2017), and Andrew Crane-Droesch, "The White House Didn't Like My Agency's Research. So It Sent Us to Missouri," *The Washington Post* (October 21, 2019).

54 For a discussion and relevant references, see Adam Rome, "'Give Earth a Chance': The Environmental Movement and the Sixties," *The Journal of American History* (September 2003).

55 Elizabeth Kolbert, "Under Water: Can Engineers Save Louisiana's Disappearing Coast?" *The New Yorker* (April 2019), 45.

56 See, for example, Peter J. Balint, Ronald E. Steward, A. Desai, and Lawrence C. Walters, *Wicked Environmental Problems: Managing Uncertainly and Conflict* (Washington, DC: Island Press, 2011).

57 See the essays in Cronon, ed., *Uncommon Ground*.

58 Bronson W. Griscom, J. Adams, Peter W. Willis, Richard A. Houghton, G. Lomax, Daniela A. Miteva, William H. Schlesinger, D. Schoch, Juha V. Siidamaki, Pete Smith, Peter Woodbury, Chris Zganjar, Allen Blackman, Joao Campari, Richard T. Conant, Christopher Delgado, Patricia Elias, Trisha Gopalakrishna, Marisa R. Hamsik, Mario Herrero, Joseph Kiesecker, Emily Landis, Lars Laestadius, Sara M. Leavitt, Susan Minnemeyer, Stephen Polasky, Peter Potapov, Francis E. Putz, Jonathan Sanderman, Marcel Silvius, Eva Wollenberg, and Joseph Forgione, "Natural Climate Solutions," *Proceedings of the National Academy of Sciences of the United States of America* 114(44) (October 31, 2017).

59 See the large-scale environmental DNA research of the sort recently funded by the National Science Foundation for work in coastal Maine. https://umaine.edu/news/blog/2019/08/21/20-million-grant-awarded-for-maine-environmental-dna-initiative-to-support-coastal-ecosystems/, and celebrated by the Forest Service, Pacific Northwest Research Station. www.fs.fed.us/inside-fs/delivering-mission/sustain/pnw-research-station-scientists-open-new-horizons-edna.

60 Peter Wohlleben, *The Hidden Life of Trees: What They Feel, How They Communicate—Discoveries from a Secret World* (Greystone Books, 2016).

61 S. Simard, D.A. Perry, and M.D. Jones, et al., "Net Transfer of Carbon between Tree Species with Shared Ectomycorrhizal Fungi," *Nature* 338:579–582 (1997). Suzanne Simard, *Finding the Mother Tree: Discovering the Wisdom of the Forest* (New York, NY: Alfred A. Knopf, 2021). See also Global Tree Search at https://tools.bgci.org/global_tree_search.php, and B.S. Steidinger, et al., "Climatic Controls of Decomposition Drive the Global Biogeography of Forest-Tree Symbioses," *Nature* 569 (May 16, 2019).

62 Steidinger, et al., "Climatic Controls of Decomposition Drive the Global Biogeography of Forest-Tree Symbioses," 404.

63 The benefits of avoiding mosquitos are manifest. See Timothy C. Winegard, *The Mosquito: A Human History of our Deadliest Predator* (New York, NY: Dutton, 2019). On the benefits of forest bathing, see Qing Li, "Effect of Forest Bathing Trips on Human Immune Function," *Environmental Health and Preventative Medicine* 15:1 (2010). See also The Japanese Society of Forest Medicine http://forest-medicine.com.

64 I have been referencing Wohlleben's book in this section. On dirt, see David R. Montgomery, *Dirt: The Erosion of Civilizations* (Berkeley, CA: University of California Press, 2007). On the science of soils, see Matt Busse, Christian P. Giardina, Dave M. Morris, and Deborah S. Page-Dumroese, eds. *Global Change and Forest Soils: Cultivating Stewardship of a Finite Nature Resource* (Amsterdam: Elsevier, 2019). This is a fairly comprehensive overview. One sentence in the introduction, pulled from a chapter, jumped out at me. "The Desire to Restore and Repair Degraded Landscapes Is an Innate Human Trait," 4, 260. I find it hard, however, to imagine any selective pressure that would have evolved such an innate human desire.

65 On the effects of the Little Ice Age in Europe, see Philipp Blom, *Nature's Mutiny: How the Little Ice Age of the Long Seventeenth Century Transformed the West and Shaped the Present* (New York, NY: W. W. Norton and Company, 2019), and Brian Fagan, *The Little Ice Age: How Climate Made History, 1300-1850* (New York, NY: Basic Books, 2000).

66 For a systematic review of the evidence and arguments, see A. Koch, C. Brierley, Mark M. Maslin, and Simon L. Lewis, "Earth System Impacts of European Arrival and the Great Dying in the Americas after 1492," *Quaternary Science Review* 207:13–36 (2019). William F. Ruddiman first argued, in "The Anthropogenic Greenhouse Era Began Thousands of Years Ago," *Climate Change* 61:261–293 (2003), that with those deaths and consequent reforestation, a substantial amount of atmospheric carbon dioxide was sequestered in new forest growth, which then contributed to a lowering of global temperatures during the Little Ice Age. He followed with *Plows, Plagues, and Petroleum: How Humans Took Control of Climate* (Princeton, NJ: Princeton University Press, 2005), in which he extended his argument. These two publications generated a robust research response, as scholars sorted out population sizes, disease fatalities, the extent of pre-contact human-modified landscapes, and the physics, chemistry, and biology of atmospheric carbon sequestration—all of which may have had implications for pre-industrial global climate change. See also Richard J. Nevle and Dennis K. Bird, "Effects of Syn-Pandemic Fire Reduction and Reforestation in the Tropical Americas on Atmospheric CO_2 during European Conquest," *Palaeogeography, Palaeoclimatology, Palaeoecology* 264:25–38 (2008); Robert A. Dull, Richard J. Nevle, William I. Woods, Dennis K. Bird, S. Avnery, and William M. Denevan, "The Columbian Encounter and the Little Ice Age: Abrupt Land Use Change, Fire, and Greenhouse Forcing," *Annals of the Association of American Geographers* 100:755–771 (2010); R.J. Nevle, D.K. Bird, W.F. Ruddiman, and R.A. Dull, "Neotropical Human-Landscape Interactions, Fire, and Atmospheric CO_2 during European Conquest," *The Holocene* 21:853–864 (2011).

67 Koch, et al., "Earth System Impacts of European Arrival and the Great Dying in the Americas after 1492."

68 Ibid.

69 G. Churkina, A. Organschi, C.P.O. Reyer, et al., "Buildings as a Global Carbon Sink," *Nat Sustain* (2020). https://doi.org/10.1038/s41893-019-0462-4.

70 On the nature of cities, see generally www.thenatureofcities.com/. Extinction, when human-caused, constitutes a failure to think ecologically. On extinction, and the loss to our planet, see Elizabeth Kolbert, "What We Lose When Animals Go Extinct," *National Geographic* (October 2019). Edward O. Wilson continues to write on this topic. See, for example, *Half-Earth: Our Planet's Fight for Life* (New York, NY: W. W. Norton & Company, 2016).

71 Kayla Epstein and Juliet Eilperin, "Greta Thunberg Had One Question at the U.N. Climate Summit: 'How Dare You?'" *The Washington Post* (September 23, 2019). As Thunberg said of Trump: "Why Should I Waste Time Talking to Him When He, of Course, Is Not Going to Listen to Me?"

72 See, generally, Preet Bharara, Christine Todd Whitman, Mike Castle, Christopher Edley, Jr., Chuck Hagel, David Iglesias, Amy Comstock Rick, and Donald B. Verrilli, Jr., National Task Force on Rule of Law & Democracy, "Proposals for Reform, Volume II" (Brennan Center for Justice, NYU School of Law, 2019). www.brennancenter.org/our-work/policy-solutions/proposals-reform-volume-ii-national-task-force-rule-law-democracy.

73 Dino Grandoni and Juliet Eilperin, "Interior Dept. Officials Downplayed Federal Wildlife Experts' Concerns about Trump's Border Wall, Documents Show," *Washington Post* (December 10, 2018). www.washingtonpost.com/energy-environment/2018/12/11/interior-officials-downplayed-federal-wildlife-experts-concerns-about-trumps-border-wall-documents-show.

74 Colin Tudge, *The Tree: A Natural History of What Trees Are, How They Live, and Why They Matter* (New York, NY: Three Rivers Press, 2005), 368.

75 See, as a few examples among many, Mike Spies and J. David McSwane, "Inside the Trump Administration's Chaotic Dismantling of the Federal Land Agency," *ProPublica* (September 20, 2019), David Shepardson, "Trump Administration Bars California from Requiring Cleaner Cars," *Reuters* (September 19, 2019), Ben Guarino, "USDA Science Agencies' Relocation May Have Violated Law, Inspector General Report Says," *The Washington Post* (August 6, 2019), and Ben Guarino, "USDA Relocation Has Delayed Key Studies and Millions in Funding, Employees Say," *The Washington Post* (October 2, 2019).

76 Oliver Milman, "Trump Administration's War on Science Has Hit 'Crisis Point', Experts Warn," *The Guardian* (October 3, 2019).

77 Oddly enough, the August 2019 U.S. Department of Agriculture Forest Service Comprehensive Capital Improvement Plan touts the Research and Development branch of the Forest Service in developing techniques of fire suppression, apparently oblivious to the ecological consequences of such suppression.

> [W]ildfire research conducted by FS R&D has led to the development of innovative firefighting technologies that mitigate the harmful effects of wildfire smoke and, as of 2013, enable the suppression of 98 percent of forest fires within the first 24 hours.

Chief Forester Silcox, with his 10 a.m. policy, would have approved.

2

WHEN THE WELL RUNS DRY

Aquifers, Canals, and the Colorado River System

> When all the rivers are used, when all the creeks in the ravines, when all the brooks, when all the springs are used, when all the reservoirs along the streams are used, when all the canyon waters are taken up, when all the artesian waters are taken up, and when all the wells are sunk or dug, there is still not sufficient water to irrigate all this arid region.
>
> —*John Wesley Powell, Irrigation Congress, Los Angeles, 1893*

The history of water development in the American West parallels the history of U.S. forestry. Over time, both rivers and forests became thoroughly entangled with government bureaucracies, within which an enthusiasm for big ideas, big conservation, big engineering, and skilled sloganeering were central components. Impounding rivers and cutting trees for human use presumably required the best science of the day to support enormous efforts of landscape modification. And yet, in a familiar story, bureaucrats in charge of water development and forest production overlooked scientific determinations that contradicted the desires of politicians and policymakers.

In the U.S. Forest Service, wildfire science was ignored. In the Bureau of Reclamation, responsible for building massive dams, hydrological science was ignored. The timelines were identical. The U.S. Forest Service suppressed fires on forests in the early to mid-twentieth century, resulting in contemporary forests prone to catastrophic fires, while the Bureau of Reclamation built huge dams and impounded massive amounts of water, resulting in contemporary allocations of water unable to satisfy human needs.

For the U.S. Forest Service, as for the Bureau of Reclamation, an imagined world of sustainable trees and perpetual water failed to match reality. Scientific information that was available to both agencies could have improved outcomes. Yet influential bureaucrats turned a blind eye toward that information. They relied

DOI: 10.4324/9781003297444-4

instead on a domain of science whose flawed and incomplete work supported agency desires. Those agency desires, conjoined with the bureaucratic power to implement decisions, became imperiled forests, which for a time supported the building industry; and they became massive public water works, which for a time supported the growth of desert cities. The pairing of political desires and bureaucratic power often goes a long way, despite abundant evidence that these desires and that power should be called into question. Perhaps only in retrospect are our enemy ancestors revealed. By then it may be too late to reverse course.

As with all cities, the history of Tucson in southern Arizona is one of increasing environmental domination. Water is the key factor: finding it, building reservoirs, canals, and pipelines for it, pumping it from deep underground and, when all else fails, filing lawsuits over it. Without water, Tucson would be a mere cluster of houses secreted in the desert with no commercial agriculture to speak of and little if any mining. Without a source of water, Tucson's golf courses, symbols of desert hubris, would shrivel up beneath the hot summer sun and blow away. Swimming pools would be useful only for skateboards and inline skates. Retired snowbirds who travel from northern climes to overwinter in the warmth of the Sonoran Desert would find more favorable places to alight. The U.S. Air Force would certainly not have a base nearby.

The Santa Cruz River used to flow year-round in the area, with periodic drought-induced interruptions. A small river, its flow was sufficient for the needs of a few thousand people, when combined with water from a handful of local springs. At one time, the Santa Cruz and other smaller rivers supported riparian zones thick with cottonwood trees and willows. Water tables were high enough to sustain large mesquite forests, called *bosques* in Spanish. Beaver, muskrat, fish, and turkey were well adapted for life in and along the desert's rivers.[1]

In the late seventeenth century, Father Eusebio Francisco Kino, a Jesuit priest who established missions in northern Mexico, thought that upwards of 5,000 people could live in the Tucson area. Long before the Spanish arrived, the Hohokam had practiced irrigation and floodwater agriculture for some 2,000 years then abandoned the region for reasons that are still murky but probably had to do with increasing temperature and aridity or, more likely, devastating floods coupled with the increased salt content of their soils resulting from millennia of irrigation. Today in the Tucson area there are more than 980,000 people. The cottonwood forests and mesquite bosques are gone, victims of the demise of the basin's rivers. Overall biodiversity is in decline, while the replacement of native species with non-native species continues apace, with some 380 alien species well established in the Sonoran Desert.[2]

In 2000, I moved to Tucson to join an international research project studying the relationships between local cultural values and national environmental policies. Part of my research interest was a desire to understand the problems of sustainable development in the southwest. How can a rapidly growing city such as Tucson survive in a desert environment? What factors will ultimately limit growth? Water, as many researchers have pointed out, is the essential resource.[3]

Water is a key problem for people and governments worldwide. Some projections indicate that by 2030, global renewable water supplies will be overdrawn by 2,680 cubic kilometers (643 cubic miles) annually. The effects will be disproportionately felt across the planet as climate changes and populations grow. Recent water crises have affected Chennai, India, São Paulo, Brazil, Cape Town, South Africa, and Los Angeles, California.[4] In North America, water supplies in the desert southwest have been a concern at least since John Wesley Powell's publication of his 1878 *Report on the Lands of the Arid Region*. That longstanding concern is compounded with global warming. It is also confounded by those who deny the reality of our changing climate.

As I began my own research, I was not surprised to learn that water consumption and use in the southwest is considerably higher than other areas. The national average in the United States is 40 gallons per person per day (1 U.S. gallon equals 3.79 liters). This included water for drinking, bathing, flushing toilets, running dishwashers, watering plants, and so on. Desert life required much more.

In the late twentieth century, people in Phoenix and Las Vegas used on average more than 300 gallons per person per day. This was seven times the national average. Swimming pools, evaporative coolers, misters, and flood-water irrigation of lawns accounted for part of the higher use.

Tucson's averages were better than those of Phoenix and Las Vegas and stood between 106 and 148 gallons per person per day. This was down from a high of about 205 gallons per person per day in the early 1970s. By 2015, Tucsonans averaged 80 gallons per person per day. These figures demonstrate a commitment to water conservation, but they also indicate a problem looming in Tucson's future. If the city doubles in population over the next 25 years or so, as some projections have it, how can it continue to function with high rates of water consumption? Where will the water come from?[5]

Groundwater

As luck may have it, beneath the Sonoran Desert lie huge aquifers, estimated to contain about 63 million acre-feet of water in the Tucson Basin and the nearby Avra Valley Basin (one acre-foot equals 325,851 gallons or 1,233 cubic meters). An acre-foot is a measure invented by the U.S. Geological Survey in the late-nineteenth century to describe the amount of water needed to irrigate an acre of land. If all of the water in the Tucson and Avra Basins was recoverable, it could, under the Geological Survey's definition, irrigate 63 million acres of farmland. Extracting water from these aquifers began in the 1870s. Windmills provided the best pumping technology but could only raise water 25 feet or so. They were often idled by lack of wind. In 1889, wood-burning steam engines provided the next technological improvement, capable of drawing 1,250 gallons a minute from 40 feet below the surface. In 1914, gas and electric pumps were introduced, at a time when rivers in Tucson still had regular flows. By the 1940s, enough water had been pumped out of the aquifers, most of it for agriculture, to lower the water table significantly, in some places by as much as 200 feet. Pumping had the effect of removing what little surface flow remained in local rivers. Riparian zones that depended upon year-round water sources died. Dry washes remained behind, on the banks of which the City of Tucson eventually built wonderful narrow parks for runners, bicyclists, and lovers of dry washes. In southern Arizona, the Santa Cruz River effectively disappeared because of an increasing, and increasingly water-consuming human population.

Barring some radical change in climate, all future river flows will be sporadic flood events, dramatic, spectacular, short-lived, with the occasional kayaker playing illegally on the waves. There is one exception: a nine-mile stretch of the Santa Cruz has become perennial once again, flowing with effluent water discharged from a wastewater treatment plant at a level no kayaker would deign to boat.[6]

Water pumped out of the ground is insufficient to meet the projected needs of this rapidly growing desert city. Geological and hydrological reports indicate that much more water is being removed than is being naturally replenished, by a factor of two. Eventually, the aquifers will be pumped effectively dry, as is happening in the much larger Ogallala aquifer beneath Colorado, Kansas, Oklahoma, New Mexico, and Texas. Since 1940, when pumping began in earnest, between six and eight million acre-feet of the most easily accessible water in the aquifers has been removed. As the water table drops, it takes more energy to lift water from the depths. Pumping costs will in time escalate. In 1983, it cost approximately $138.10 to pump a million gallons of water (about 3.068 acre-feet). In 2000, the average cost to pump a million gallons of water for both gas and electric wells was still fairly modest, $162.05. These figures are only a small portion of the overall costs needed to deliver water to consumers, which also include administrative, distribution, and capital repayment costs. In the early 1980s, several scholars argued that such energy costs, no matter how high, would not matter at all in two or three generations, when the groundwater may be gone, given the projected rates of consumption. By 2000, the City of Tucson, mostly through education and conservation programs, had slowed groundwater removal, but not enough to be sustainable.[7]

One effect of removing groundwater is subsidence. The earth above the aquifer compacts, sometimes a few inches, sometimes several feet. The ground no longer absorbs water easily. Cracks and fissures appear on the surface. More than 3,000 square miles of Arizona have subsided. As water is mined, the overall elevation of Arizona is thus lowered. Projections for as much as 12 feet of subsidence have been made for areas around downtown Tucson. At the 2000 rate at which water was sucked from the ground it may take fewer than 25 years for such remarkable drops of elevation to occur. A basketball rim stands at ten feet. When reading the geological reports, I have thought that the City of Tucson might take advantage of the situation and build sunken basketball courts, a cooler place for kids to play hoops. Subsidence of course is not unique to Arizona. It has occurred in Mexico City, Beijing, and Tokyo, among other places.[8]

The U.S. Federal Government, the State of Arizona, Pima County, and the City of Tucson are aware of the various problems and took some steps to remedy them. In 1980, for example, under pressure from Interior Secretary Cecil Andrus, a comprehensive Arizona Groundwater Management Act was passed, which created the Arizona Department of Water Resources. The basic goal of the Department is to ensure that water for the State does not run out, which means that groundwater removal must at some point equal groundwater renewal. The critical areas are the population centers of Prescott, Phoenix, and Tucson, the agricultural area of the Pinal water management district, which lies between Phoenix and Tucson, and a district called Santa Cruz, which encompasses the city of Nogales on the Mexican border. For Prescott, Phoenix, and Tucson, the safe-yield goal is set for the year 2025, when supply should equal demand.[9]

As the Department's reports make clear, however, increasing population and new industry may render this goal impossible to meet. After 2025, if the projected Tucson population of 1.25–1.6 million continues to grow, all bets are off. Even if a balanced water budget is met by 2025, the results may be extreme, with trade-offs between human consumption and landscape. As one research report noted in 1988: "the effects on individuals' lifestyles and on the total environment would be severe. Essentially, all greenery within the metropolitan area would disappear."[10] The loss of greenery may be a dire prediction, offset by recent practices and ordinances requiring use of indigenous drought tolerant plants. Still, if greenery decreases because of water conservation measures, then the buildings and streets of Tucson would absorb even more energy from the sun, further raising the temperature of the city, which in many places is already two degrees Celsius higher than surrounding rural areas. Coupled with global warming, Tucson could become hot indeed.[11]

For the Santa Cruz district to the south of Tucson, the goal is simply to keep water tables from dropping at all and to maintain the present safe-yield level. The Pinal water goals to the north are much more ambiguous and resemble the disastrous use of the Ogallala aquifer. These goals are, as the management plan has it, "to protect the agricultural economy as long as feasible, and preserve water supplies for future non-agricultural purposes." To protect the agricultural economy means a planned depletion of Pinal groundwater, which since 1948 has meant pumping more than 43 million acre-feet from the underlying aquifer. The ultimate equation is quite simple. When the water is gone, commercial agriculture stops. In other words, grow crops until the aquifer runs dry. The little remaining protected water, for future, unspecified non-agricultural uses, lies between 1,000 and 1,200 feet below the surface.[12]

Colorado River Water

For much of the second half of the twentieth century, local residents have refrained from addressing the threat to continued human habitation in the area because of the promise of renewable water brought in from the Colorado River. Farmers, politicians, and Tucson's residents apparently believed that Colorado River water will keep underground sources from being depleted. Water from this distant source would allow Tucson to prosper, in part by stemming the flow of precious groundwater into Tucson's faucets, toilets, and swimming pools. Tucson's alternative newspaper, *The Tucson Weekly*, was the single dissenting public voice and frequently published informed diatribes against the use of Colorado River water and its presumed potential to alleviate local water problems. "Pumping Money," "Pumping Bile," and "CAP is still Crap," proclaimed the headlines.[13]

Transporting Colorado River water to central and southern Arizona has not been easy. As a political problem, developing Colorado River water for use else-where had its origins in the early twentieth century. The first solution was to create the 1922 Colorado River Compact, which arbitrarily divided the Colorado River watershed into upper and lower basins. Arizona, California, and Nevada make up the lower basin; Colorado, New Mexico, Utah and Wyoming make up the upper basin. Small portions of Arizona and New Mexico were placed in both basins (Figure 2.1).[14]

Geographically, the dividing line between the two basins is at Lee's Ferry, a wide shallow spot on the Colorado River where it is joined by the Paria River in nor-thern Arizona before it enters the Grand Canyon. Named after John Doyle Lee, a Mormon who homesteaded the area at Brigham Young's suggestion, Lee's Ferry seems an appropriate historical site to divide western water, even if the geographical division is questionable.

John Lee, along with other Mormons and a group of Paiute Indians, had participated in the infamous 1857 Mountain Meadow massacre of 120 men, women, and adolescent immigrants headed to California from Missouri and Arkansas. At the time, the U.S. government had sent troops to Utah and Mormons were edgy about federal incursion into their lands. What triggered the killing, however, was trivial. A few taunts and slurs about Mormons, perhaps about their marriage practices— Lee had 18 wives—tossed off by members of the immigrant wagon train passing through The State of Deseret, as the Mormons called their territory. Horrified by the killing, Mormon leaders tried to hide the massacre and excommunicated Lee and others from the church. They suggested Lee move deep into the desert where he could nonetheless continue to serve the church from afar. Twenty years later, in 1877, Lee was captured and executed for his crime. He was the only one of his party to meet this fate, "clearly a scapegoat for a wider guilt," as historian Donald Worster has noted.[15]

It is historically fitting to divvy up western water at Lee's Ferry, where many of the contradictions of western life are so starkly apparent. This is a region marked by a strong sense of individualism, religious nation-building in a country that insists on the separation of church and state, Indian-white conflict and cooperation, and polygamous murderers hiding out in the desert—the stuff of western history and myth. Federal power was eventually to play out within the vastness of the Colorado River drainage. Here, after the turn of the century, the wider guilt Worster speaks of took new forms, and new contradictions became apparent, with the Colorado River and Lee's Ferry playing increasingly central roles.

Those who controlled water in the arid West controlled the West's destiny. States, fearing for their futures, began to fight for a share of Colorado water. The architects of the Colorado River Compact were determined to devise a rational and federally

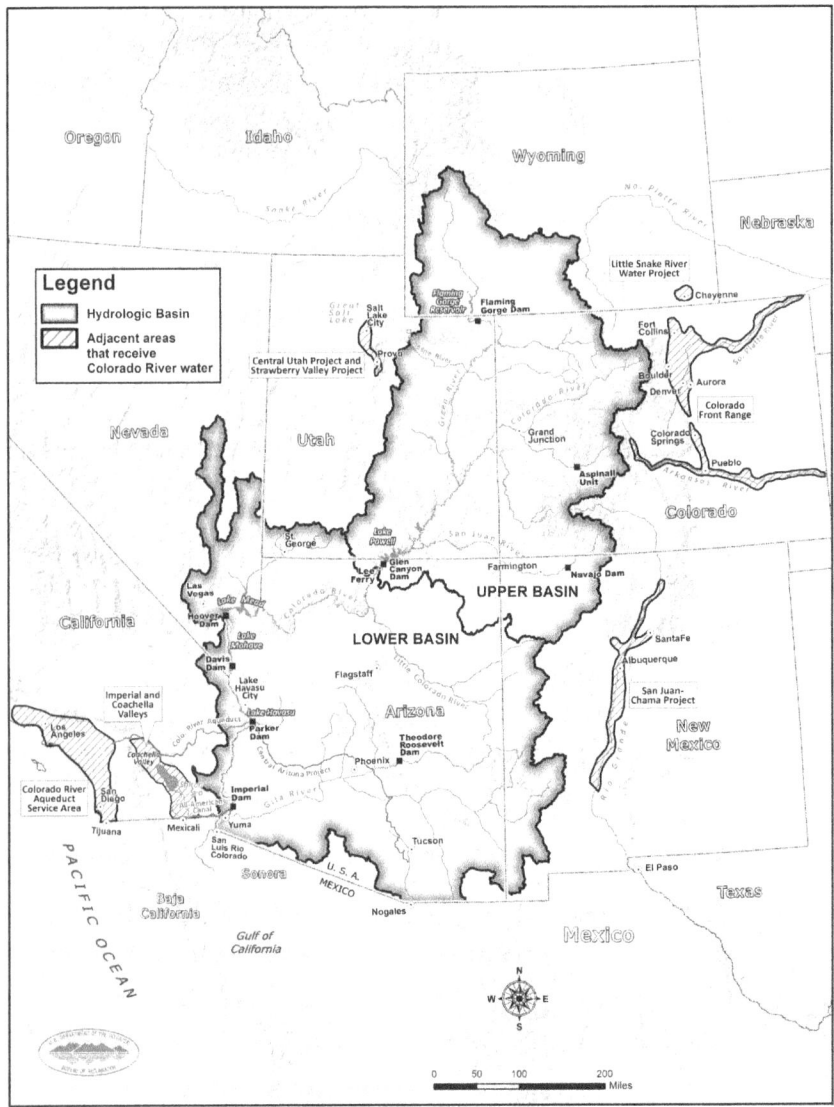

FIGURE 2.1 Map of the Colorado River

Source: Bureau of Reclamation, U.S. Department of the Interior. www.usbr.gov/dcp/.

mandated plan to provide water for western development, and to allocate water among squabbling states for all future uses. In 1928, after six years of stalemate and consequent Congressional intervention, six of the states signed the compact. That same year the Boulder Canyon Project Act apportioned Colorado River water to the lower basin states. Nevada was allocated 300,000 acre-feet, California 4.4 million acre-feet, and Arizona 2.8 million acre-feet. Believing its water needs were being slighted, the State of Arizona—the lone, petulant holdout—refused to sign the compact until 1944, the same year that the Mexican Water Treaty committed the United States to deliver 1.5 million acre-feet of Colorado River water to Mexico.[16]

Under the terms of the compact, each basin got 7.5 million acre-feet per year to apportion among its member states, based on vastly over-inflated flow estimates, some as high as 22 million acre-feet per year. The compact used a 16.8 million acre-feet per year figure, calculated from measurements taken near Lee's Ferry, but, as later researchers discovered, this estimate was derived from an unusually wet ten-year period between 1914 and 1923. A more accurate flow estimate, calculated from tree-ring data over a 400-year span, puts the long-term yearly average at 13.5 million acre-feet per year. But the initial Lee's Ferry estimate would nonetheless hold, over-committing available water based on inaccurate flow estimates and defining future relationships between states. The result was a political tradition in the arid West that ignored empirical constraints.[17]

With the Colorado River Compact finally in place, the second political solution to bring Colorado River water to central and southern Arizona was the creation of the Central Arizona Project (CAP), wedded for a time to the ill-fated dams proposed for the Grand Canyon.

The Grand Canyon dams were to produce hydroelectricity, the sale of which would pay for a scheme to add water to the Colorado River. According to this absurd scenario, water would come from the Pacific Northwest, from the Columbia River, which carries ten times as much water as the Colorado. After over-committing the water nature provided, the bureaucratic vision was not to scale back growth in the arid West, nor was it to forge a human relationship to the Colorado River that was proportional to its size. The solution was to construct a system to transport Columbia River water to the Colorado River, and thereby increase the size of the Colorado to bring it in line with inaccurate flow measurements: Nature-by-design, or by government fiat.[18]

Environmental groups rallied to defeat the proposed dams in the Grand Canyon, but compromised on the location of Glen Canyon Dam, which blocked the Colorado River and formed Lake Powell. It is worth pausing a moment to consider the history of that lake. Completed in 1966, Glen Canyon Dam nearly gave way in 1983. A wet fall in 1982, a heavy winter snowpack, and an unusually warm spring combined to make runoff in the Colorado River watershed considerably higher than average. In January of 1983, Lake Powell was 90 percent full. By the end of May, the reservoir was 96 percent full. As snowmelt flowed into the lake at 90,000 cubic feet per second in June, it continued to fill beyond its capacity. Flows rose to over 120,000 cubic feet per second by July. The dam, however, had been designed to release less than half that amount through its turbines.

Two spillway tunnels carved through the sandstone adjacent to the dam—safety valves for an overfull reservoir—gave some measure of relief to dam operators. Their use almost immediately presented a problem, however. The massive release of water caused the spillways to begin to fail. Concrete linings of the spillway tunnels fell apart under the rush of water, with enough debris in one instance to clog a tunnel and block the flow of water. Engineers were afraid that once the linings failed, water would quickly erode the soft sandstone and potentially undermine the integrity of the dam. If the dam failed, the result would be a cascading disaster, as dams below Lake Powell collapsed in turn under the onslaught of a massive flood. Hoover Dam (built 1931–1935), which formed Lake Mead, and Parker Dam (built 1932–1936), which formed Lake Havasu, were at risk as well. Engineers solved the problem by securing sheets of plywood and then metal panels to the top of the spillway gates, which allowed the lake to rise higher, within seven feet of the top of the dam. They then waited for the lake level to drop below the spillway height, as runoff slowed to manageable levels at the end of the summer.

In 1983, the problem was too much water. In the early twenty-first century, the problem is too little water, with sustained drought a prominent concern for cities and farmers alike. As James Lawrence Powell emphasizes in his book *Dead Pool*, "hydraulic societies have always deluded themselves into believing they could gamble with Nature and win." As with all such gambles, humans typically manage to find short-term success. In the long term, however, the gamble will fail. The question that remains is not whether but when Glen Canyon Dam and Lake Powell will succumb to the natural law of the river. With 37,000 acre-feet of silt settling onto the floor of Lake Powell each year, the reservoir will eventually fill with enough mud to render Glen Canyon Dam useless—not for 700 years, in the Bureau of Reclamation's estimate; in as few as 51 years, in one worst-case scenario. Unfortunately, global warming and reduced runoff accelerate the process.[19]

The Central Arizona Project

Despite the uncertainties of water flows, dams, and climate in sustaining cities in desert environments, the Central Arizona Project (CAP) received strong political support. Secretary of the Department of the Interior Steward Udall and his brother Arizona Representative Morris Udall—grandsons of John Doyle Lee—were key political figures in keeping the CAP going, as was Arizona Senator Carl Hayden, for whom the project was a long-term goal. It was finally authorized in conjunction with many other water projects under the Colorado River Basin Project Act and signed into law by President Lyndon Johnson in 1968. After five more years of political haggling, construction began. The economic benefits of hydroelectric dams within the Grand Canyon, the environmental costs of filling the Grand Canyon, and the engineering problems with transporting Columbia River water to the Colorado were no longer under consideration. In this instance, values associated with environmental preservation seemed to trump values associated with massive public works projects.[20]

CAP water was initially intended to expand Arizona agricultural lands, in the hope that even more desert with its long growing season could be converted into farmland. But by 1968, it had become water for agricultural salvage. So much water had been pumped out of the ground in central Arizona that no new agricultural lands could be developed. CAP water was needed to sustain the lands already under cultivation. Moreover, the authorization act required that for every acre-foot of delivered CAP water, one less acre-foot of groundwater could be mined. There was a further problem. By the late 1970s, it was clear that growing populations in Phoenix and Tucson would need CAP water for residential not agricultural purposes. As these cities expanded, they encroached upon former farmlands and converted them to suburbs; CAP water would eventually be used to irrigate suburban lawns. In addition, under the 1908 Supreme Court ruling Winters v. United States, Indian tribes on reservations were entitled to a share of water at levels sufficient to support their communities—but the percentages, more than a century later, have yet to be determined for many of them.[21]

The political problems were nearly intractable. California's congressional delegation stalled the CAP as long as possible because in the interim southern California could use all of the water that would otherwise go to Phoenix and Tucson. To finally win Congressional authorization of the project required, moreover, the so-called California Guarantee. This guarantee meant that California received all its allocated water before Arizona could take any. If California insists on this provision during a prolonged drought, little if any Colorado River water will be pumped to Phoenix and Tucson. Still, political compromises in place, CAP construction began in 1973 with high hopes for the future growth of metropolitan Arizona.[22]

By 1977, the Central Arizona Project was well underway, yet increasingly excessive costs did not equal potential benefits and President Jimmy Carter, in an attempt to save more than $9 billion in federal money, announced that the project, along with 19 other similar water reclamation efforts, would be halted. Carter eventually changed his mind under intense pressure from western politicians who, as it turns out, were less fiscally and environmentally prescient than the former president. Many of them had staked their political careers on bringing federally funded water to arid western cities, and they did not appreciate efforts to undermine their desires. Everyone seemed willing to ignore the science of the day. Table pounding in committee meetings substituted for rational environmental discourse, as western politicians gathered support for their water works, and prevailed. The Central Arizona Project would continue.[23]

While the political problems were difficult, the engineering problems were easier; they simply needed lots of money.

Transporting Colorado River water meant lifting water 2,900 feet from Lake Havasu, constructing 14 pumping stations, and digging 336 miles of canals and tunnels to reach the south side of Tucson. Much of the power for such heavy lifting came from the coal-burning Navajo Generating Station near the town of Page in northern Arizona. The Bureau of Reclamation, which built the CAP, bought nearly 25 percent of the Navajo Station in order to have sufficient power to move Colorado River water south and east. Coal for the plant was strip-mined on Black Mesa, transported 75 miles to the generating plant, burned to produce steam to move turbines to produce electricity, in order to move relatively recent snowmelt several hundred miles from the Colorado River Drainage System deep into the Sonoran Desert where it rarely snows. All of the effort, from lifting water 2,900 feet to diverting it in 336 miles of canals, was expended so that the Arizona metropolises of Phoenix and Tucson would stop depleting their precious groundwater.[24]

Black Mesa sits on Navajo and Hopi reservations. The Peabody Western Coal Company leased rights to the coal from the Navajos and Hopis and also negotiated an arrangement to use Navajo water to transport the coal by slurry to another power station, the Mojave Generating Station, some 270 miles distant in Nevada. Richard White, in his analysis of the modern rise of the metropolitan West, describes the results:

> The lease provided the coal at prices well under its market value, and it virtually gave away the precious water. Another Interior Department agency, the Bureau of Indian Affairs, had a trust obligation to protect Navajo and Hopi interests, but it approved the contracts. This was how growth worked. Indian energy and water subsidized Phoenix's [and Tucson's] energy and water. The Indians lost; Peabody Coal and the metropolitan West won.[25]

The win was short-lived. In 2019, after more than 45 years of power production, the Navajo Generating Station shut down. Over its life, the power plant was a huge carbon emitter, releasing nearly 135 million metric tons of carbon dioxide between 2010 and 2017. Despite pressure from various stakeholders, including the Bureau of Reclamation, which needed power to move CAP water uphill to Tucson, the plant was closed for economic and environmental reasons.

Meanwhile, power to move water across the desert had become cheaper. In 2016, CAP bought $81.2 million worth of Navajo Generating Station power and resold $12 million as surplus. Power on the open market, however, would have cost $42.7 million, potentially saving CAP $38.5 million. It finally made sense to the Bureau of Reclamation to agree to close the Navajo Generating Station and for CAP to buy power on the open market.[26]

With the closing of the power plant, the Black Mesa mine supplying coal also shut down. Markets for coal were in decline, while markets for natural gas were in ascension. Coal was no longer an option. The economic effects of the mine closure, however, have proven difficult for Hopi and Navajo tribes. They had come to rely on royalty revenues from the coal. For the Hopi, 80 to 85 percent of general fund budget monies came from coal royalties—some $12 million annually. For the Navajo, the loss was estimated at between $30 and $50 million in coal revenue for 2020. Both tribes struggled with those losses.[27]

By the standards of western water projects, where the term "boondoggle" is too often appropriate, the CAP is on the large side. Begun in 1973 and substantially completed in 1993, it has the capacity to bring about 1.5 million acre-feet of water each year into central and southern Arizona. Construction costs were $4.7 billion. Marc Reisner, whose book *Cadillac Desert* details the sorry history of water allocation in the West, provides an apt characterization of the project: "as incongruous a spectacle as any on earth: a man-made river flowing uphill in a place of almost no rain."[28]

The desert river that flows uphill made little economic sense, as Maurice Kelso, William Martin and Lawrence Mack pointed out in 1973 in their book *Water Supplies and Economic Growth in an Arid Environment*. But by 1993, it was an accomplished fact. The result, at least for Tucson, was not what most people expected, bureaucrats and politicians among them.

By 1995, CAP water was banned for residential use by voter initiative because it tasted bad, had an offensive odor, and contained high mineral levels that corroded old pipes. Few Tucsonans wanted to drink it. The water damaged their dishwashers, evaporative air conditioners, and water heaters, among many other water-dependent appliances. Fish in aquariums died. Houseplants wilted. Pipes sprung leaks. Tucsonans were unhappy. The City of Tucson paid out $1.9 million worth of CAP related claims to some 5,300 citizens, apparently a small fraction of the damage caused by the water brought from afar.[29]

That the water from far away turned out to be corrosive and bad-tasting was an extraordinarily unfortunate turn of events. Was it unforeseen? After billions of dollars, dozens of lawsuits, several decades of persistent political maneuvering at state and federal levels, and 20 years' anticipation as canals were dug and pumping stations built, Tucsonans simply refused to drink Colorado River water. The completion of the project, however, had already triggered the 1994 organization of the Central Arizona Water Conservation District, a state entity charged with operating CAP and repaying Arizona's obligation of $1.8–2.3 billion of the costs associated with the project. Not wanting to drink the water, Tucsonans were nevertheless obligated to pay their share of the project's costs.

"Too thick to drink, too thin to plow," the old timers used to say about the Colorado River when it ran red and muddy. One wonders what the new saying may be. Even at its source, high in the Colorado Rockies or in Wyoming's Wind River Mountains, you cannot drink the water neat. Human use of the backcountry has resulted in a dramatic increase of water-borne giardia, a one-celled organism that produces what is described as explosive diarrhea in those who have ingested it. Water filters or other treatments are needed in the backcountry. And along its course, new contaminants threatened to work their way into the river. Outside of Moab, Utah, 13 million tons of tailings from a uranium mill were situated near enough to the Colorado River to allow radioactive material and other hazardous wastes such as arsenic, lead, and mercury to drain into the river given a large enough flood. The federal government initially wanted to leave the pile of radioactive tailings where it was. In response to various lawsuits and other forms of political pressure from those downriver, the House and Senate both approved legislation to move the tailings away from the flood plain, signed into law by President Clinton on October 30, 2000.[30]

After Tucson voters expressed their displeasure over Colorado River water in the form of a Water Consumer Protection Act, the City of Tucson was required to use CAP water, which it had already contracted to buy, in only a few ways. The city could sell or exchange it for other, more desirable water. It could allow CAP water for agriculture, mining, parks, golf courses, and schools. In addition, the city could use CAP water to prevent land from subsiding and inject it into wells, but only if it was effectively treated and "free from disinfection byproducts." There was one possible exception: CAP water could be delivered as potable water if it matched in quality the groundwater Tucsonans had grown accustomed to. No mention is made in the Act of possible radioactive contamination.[31]

One strategy Tucson has actively pursued is to refill aquifers by dumping CAP water back into the ground, along with effluent from wastewater that has been treated in sewage treatment plants. Other proposed solutions included inflatable dams that could be blown up during rainstorms, plugging riverbeds long enough for captured water to percolate into the ground, after which the dams would deflate. But the CAP and wastewater recharge solutions are the ones that have been implemented.[32]

In the Avra Valley, west of Tucson, a $73 million project included the construction of large "spreading basins" into which CAP water is pumped. This water, spread over three 20-acre basins, sinks into the ground, ridding itself of impurities as it goes, losing about one and one half percent to evaporation before it settles into the earth. After about six months, some portion of this water reaches the underlying aquifers, where it blends with deeper, purer Pleistocene water, to be pumped out and used for municipal purposes.

The blended water began to reach households in May of 2001. By 2007, blended water accounted for about half of Tucson's water use. The city anticipated that up to 60,000 acre-feet of CAP water will be recharged into aquifers each year. Some wells can then be shut off, allowing levels of aquifers to again rise. It is unclear whether earth that has subsided from previous water withdrawals will readily absorb the recharged water. The overall costs of the project approached $250 million. By 2020, the blended water, dubbed "Clearwater," had become the residential norm. Some aquifer levels rose by as much as 50 feet.[33]

The belief that the Colorado River may indeed provide a sustainable source of water and a replenishment of local groundwater for the City of Tucson was short-lived. The onset of a megadrought in the American southwest shrunk snowpacks, dried soils, and limited the amount of water flowing into reservoirs. This megadrought has an anthropogenic source, at least in part, in the form of global warming. The immediate result has been a diminishment of Colorado River water to its various users, including Arizona. In 2020, Arizona, operating under a Drought Contingency Plan, reduced its annual use of Colorado River water by 192,000 acre-feet.[34]

Undrinkable Colorado River water and a diminishing supply of pure groundwater: mix them together, call the mixture "Clearwater," and the problem appears to be solved. At least in the short run and absent a continuing drought, the CAP architects may still believe in the project. Four main historical lessons seem clear enough. First, the City of Tucson is sustainable only if water is carried by canal nearly 350 miles from the northwest and dumped into basins where it will join underground aquifers after expected losses from evaporation. A perennially flowing river, however, is not enough to ensure a sustainable future for Tucson. And it is not at all clear the Colorado River will flow perennially. Most projections show a diminished river in the near future, which means diminished reservoirs, diminished water transported to Tucson, and an increasingly thirsty city.

Pumping water uphill needs water, of course, but it also needs energy. Affordable and continuous supplies of energy to pump Colorado River water into the Sonoran Desert are as uncertain as the river itself. And Black Mesa, if it reopens, will eventually be strip-mined bare.[35] If Black Mesa does not reopen, and in the event natural gas sources become costlier, new sources of cheap energy will be needed to move water such great distances.

There is a second historical lesson here. The City of Tucson is only sustainable if all taxpayers in the United States underwrite the true costs of water in the desert, as they did for the CAP project. In effect, all U.S. taxpayers have allowed Tucson to grow, partly by directly funding the CAP project and partly by allowing federal agencies such as the Bureau of Indian Affairs to agree to contracts that benefit the metropolitan West at the expense of Indian communities who have had a much longer presence in the area. Whether the result is good, environmentally sound policy is open to debate.

I tend to think it is not. But once such a project is in place, it has long-term environmental and economic consequences, and new generations will be forced to confront and adapt to the decisions of earlier generations. Our well-intentioned predecessors who mobilized with state support technologies of landscape modification may have unwittingly become enemy ancestors to their descendants. We are, after all, forced to live with the consequences of those modifications. Choices four generations ago are ours to confront. *Enemy ancestors* may well be a particularly appropriate characterization of their efforts.

The third lesson is this. Tucson is sustainable only if population growth slows dramatically. If it does not, and in truth there is no sign that it will, all the existing groundwater in the region, and all the water Arizona can squeeze out of the Colorado River, will be insufficient. That much is clear. The only solution, for which there is recent evidence, will be continued water conservation measures, for both agriculture and municipal uses. Climate change will go a long way in pushing those measures. Where they lead remains a mystery.

The fourth lesson may be the most important. We ignore science at our peril.

The Fourth Lesson

The architects of the 1922 Colorado River Compact found an advocate in Arthur Powell Davis, nephew of John Wesley Powell. Arthur Powell Davis became director of the Reclamation Service in 1914 (renamed in 1923 the "Bureau of Reclamation"). In 1920, the Kinkaid Act authorized the Reclamation Service to prepare a report on the Colorado River—resulting in the so-called "Fall-Davis" report, which provided the basis for water allocations formalized in the Colorado River Compact. Sufficient water for agriculture was the key issue; to "reclaim" desert land otherwise "wasted" was the goal. The Fall-Davis report, however, which has had long-lasting consequences, was based on faulty science and inaccurate flow measurements. It presented a rosy picture of water storage and use that the river itself could not deliver, a picture in line with visions of western development and the desires of western developers.

Eric Kuhn and John Fleck, in *Science be Dammed: How Ignoring Inconvenient Science Drained the Colorado River*, show that another report, *Colorado River and its Utilization* (USGS Water Supply Paper 395, published 1916), should have been the basis of the Colorado River Compact. Written by Eugene Clyde (E.C.) LaRue for the U.S. Geological Survey, *Colorado River and its Utilization* provided a contrasting and cautionary view of the river. As Kuhn and Fleck note, LaRue's report "described the river and its major tributaries, summarized and evaluated the available hydrology records, inventoried existing irrigation systems, identified potential irrigation opportunities, and identified and evaluated reservoir sites for river control, irrigation supply and hydroelectric power purposes." In LaRue's estimation, the Colorado River did not contain nearly enough water "to irrigate all the irrigable lands lying within the basin."[36]

Planners, bureaucrats, politicians, and developers thought otherwise. Debates between states, often based on technical legal and water allocation issues, depended on the inaccurate flow estimates of the Fall-Davis report. LaRue, however, continued to push his version of hydrological science, along with several colleagues, and he continued to conduct important field studies. Evidence that the Colorado River did not, and could not, deliver enough water to satisfy development projections was readily available and became increasingly clear with LaRue's work. And yet it was ignored.

Three main problems of the Fall-Davis report pivoted on (1) historic flow estimates of the Colorado River, (2) contemporary flow estimates that failed to account for actual use, and (3) presumptions about evaporation rates of anticipated reservoirs. Commissioners wrangling over future water use in this vast drainage assumed they had 17.3 million acre-feet to allocate. They had much less than that, however, closer to 11.3 million acre-feet once current uses and other factors were considered. LaRue pointed out the methodological and empirical flaws of the Fall-Davis report. His 1916 *Colorado River and its Utilization*, had it become the basis for allocation debates, would have led to a very different allocation scenario—one perhaps more congruent with empirical realities.

"It's clear," Kuhn and Fleck write, "the commissioners and their advisors were confused about what exactly was being measured at the different gauge locations, the policy implications of using average flows, and the problems with the different periods of record presented in the Fall-Davis report." They go on to say:

> That all this should have been confusing to them is to be expected. But their sin, which became the original sin of Colorado River Basin management, was a lack of humility in the face of their ignorance. Because while uncertainties were unavoidable given the modest nature of the available data,

they had at their disposal a team of federal scientists, led by E. C. LaRue, who had carefully thought through the implications of the uncertainties. The commissioners never asked for their help.[37]

Subsequent years, leading to the CAP with its uphill river, continued to see a similar lack of humility. The 1928 *Report of the Colorado River Board on the Boulder Canyon Project* reassessed river hydrology and concluded that an important flow measure was "10 percent too high," and, following LaRue and his colleagues, that nineteenth century drought information needed to be part of the equation. The report urged caution. It was ignored.

Later estimates chose to disregard dry years in calculating Colorado River flows and developed questionable calculations to determine "salvage by use" of water. Salvage by use, which recalls Graves' "lumbering for use," is the idea that rather than allowing evaporation or riparian vegetation to "waste" water, it is better to put that water to human purpose. The Gila River, the most important Colorado River tributary in the lower basin, had one million acre-feet of salvageable water—or so went the claim. What this idea accomplished was not so much additional water for human use, but a way for competing states to increase river flows on paper, and thus make larger demands on shares of Colorado River water.[38]

By the 1960s, scarcity, not abundance, was the message. It was clear the Colorado River did not have enough water for all the development needs in the upper and lower basins, let alone for 1.5 million acre-feet to flow into Mexico. The proposals to place dams in the Grand Canyon, and to augment Colorado River flows with Columbia River water, came and went, but the CAP continued. The 1968 Colorado River Basin Project Act saw to both, and indeed precluded the Department of the Interior from studying ways to augment the Colorado River with water from other basins. Washington senator Henry Jackson, with humor missing from our present politics, told Arizona senator Morris Udall to abandon his desire to bring water to Arizona from afar. He wrote, "According to my reading of the law, you can't *study* it, *contemplate* it, or even *dream* about it. The only thing you are permitted to do is *forget* about it." Udall, apparently, agreed.[39]

With additional water from the Columbia River a fantasy, the Central Arizona Project nonetheless increased its physical capacity, from 1.2 million acre-feet to 1.6 million acre-feet annually, by enlarging the canal system. The intent was to bring more water into Arizona during wet years, when the upper basin released surplus water downstream, and to scale back uphill transport to Phoenix and Tucson, during drought years when, presumably, it is most needed. This brings us full circle back to E.C. LaRue, whose early-twentieth-century studies of the Colorado River and its potential to support western development can be seen, retrospectively, with a clarity they should have been accorded in his time. The Colorado River did not then, and does not now, have enough water for everyone.

Culture-Nature

It is no longer possible to think of the natural world as distinct from the human world. Environmental historians have long known that natural environments, even those that appear to be unsullied by humans, are frequently creations of past human activity, at least in part. Before Europeans arrived in the Americas, for example, humans had already substantially altered ecosystems, sometimes dramatically, through hunting and fishing practices, farming techniques, widespread use of grassland and shrub-land fires, and their own social interactions. Contemporary ecologists who thought their studies were only about the natural world are beginning to recognize that human involvement must be factored into any adequate ecosystem analysis for both past and present environments. This is certainly true of water in the West.[40]

John Wesley Powell, who first floated the length of the Colorado River in 1869, suggested to the 45th Congress in his 1878 *Report on the Lands of the Arid Region* that irrigation districts should be the organizing feature of the arid West. Such districts, made up of property owned by nine or more persons, would be confined to lands that government surveyors deemed irrigable. Water, held in common, would be guaranteed for each property, which would not exceed more than 80 acres per individual. At all costs, water should be controlled by individual local farmers organized into collective water districts, which would build and maintain ditches and canals. Private entrepreneurs who would own, develop, and market water would not be allowed to operate, nor would the federal government play a major role, except as scientific advisor. Powell based his suggestion on Mormon water apportionment policies, which allocated water held in common to the benefit of all landowners. Mormons borrowed the practice of collective water control from Hispanic farmers along the Rio Grande, who in turn had incorporated Indian irrigation techniques.[41]

Had Congress adopted Powell's suggestion, one wonders what kinds of regional identities would have resulted. Defining a state as a series of water districts, rather than by the straight-line triangular logic of the surveyor, may have produced collective identities that are more ecologically sensitive than those we see today. If state boundaries coincided with natural boundaries, efforts to alter the courses of major rivers, to move water from one basin to another, and to support the rapid growth of desert cities, would have required a keener sense of natural processes, a more subtle connection between nature and culture. As it stands, the connection between nature and culture is still there, but it is blunt, unsubtle, manifest in large construction projects that link distant ecosystems and in political machinations that fund such projects.[42]

The Colorado River has become part of a vast plumbing system. It is still a natural system, dependent upon weather patterns, geological processes, and laws of physics. But it is also a cultural system, governed by dams, laws, and political relationships. Its water has been diverted, stored, and apportioned. Natural spring floods and low winter flows have been evened out, changing riverine ecology in the process. Artificial floods, intended to restore eroding sandbars in the Grand Canyon, have been tried as a substitute for natural floods, with encouraging results. Sixty non-native species of fish, introduced by federal, state, and local agencies, are well adapted to life in the dam-controlled Colorado River basin environment, and in many instances they successfully outcompete the 32 species of native fish. The river has become an "Organic Machine," to borrow the title of Richard White's book on the Columbia River: neither natural nor unnatural, but both—separating the categories makes little sense. It is more accurate to combine the categories of nature and culture, to see humans as inextricably and deeply entwined in the natural world, and to recognize all environmental issues as characterized by the contradictory relationships humans have developed with the world they inhabit. The question then becomes how best to effect the twining of nature and culture while bringing the contradictions into full view.[43]

The City of Tucson began the process of depleting its underground water sources in the late-nineteenth century. By the 1960s, it became clear that the water would eventually run out, and plans were made to capture water from afar—lower quality, unpalatable, mineral-laden water, not the pure Pleistocene water locals had grown accustomed to. At the time no one thought to ask whether Tucson residents would drink or use Colorado River water once it arrived in their taps, splashed into their bathtubs, and trickled into their washbasins. Still, not content to make use of the water resources at hand, Tucson, the State of Arizona, and the federal government ranged farther afield, and in a display of technological sophistication—or ecological arrogance, depending upon your point of view—pumped water deep into the desert, whereupon local people turned their noses up at the expensive, noisome gift.

City dwellers typically do not live lives in intimate contact with the natural world. Many of their relationships with nature are mediated by technology, by regulations governing their activities, and by the form of the city itself. That form is not self-contained. The Santa Cruz River used to flow from Tucson north into the Gila River, which in turn, before development, flowed into the Colorado River, contributing during wet years as much as one million acre-feet to that river's flow. By lowering its water table and drying up the Santa Cruz, Tucson effectively detached itself from the Colorado River drainage. With the completion of the CAP, the direction of flow has been reversed and, for good or ill, Tucson is once again within the Colorado River system, but as a recipient of rather than a contributor to the Colorado River.

Tucson's residents, refusing to drink Colorado River water unless it is blended with the sweet, ancient water beneath them, are part of the river nonetheless, at least for the foreseeable future. Tucson is thus connected in a new way to the river's tributaries, watersheds, and mountain sources, as well as to the states, Indian tribes, and other water users who claim a portion of the river as their own. What began in the nineteenth century as a simple need for water, satisfied by pumping it from below ground, has in the first decades of the twenty-first century become a cluster of needs, a web of connections, and the ecology of the Colorado River has become vastly more complex as a result.

The increasing ecological complexity of the river comes from the human side of things. Laws, political relationships, international treaties, technological improvements, bureaucratic structures, science, commerce—their successes and their failures—are now part of the ecology of the Colorado River system. Nature and culture together, river and plumbing system as they articulate or fail to articulate, provide the bases for environmental change or stability.

This is not to say that the ecosystems associated with the Colorado River drainage are in good shape. They are not. Many environments in the drainage have been seriously degraded by human activities. Native plant and animal species have become endangered, and non-native species have proved to be hardy invaders, often supplanting native species. Sediment flows which once formed a significant part of the riverine environments on the Colorado Plateau have been curtailed.

Before Hoover Dam was built, 180 million tons of silt were carried each year by the Colorado River, a sediment load that was reduced to 13 million tons when that huge plug was in place. Sediment now accumulates behind dams, the effect of which will be the eventual failure of the dams unless massive dredging projects are undertaken.

Where will the sediment be put? Who will fund such projects? Accumulating sediment also displaces water, so that the reservoirs become less efficient over time. In Lake Mead, formed by the Hoover Dam, 137,000 acre-feet of water each year are lost to silt. In Lake Powell, formed by the Glen Canyon Dam, as much as 70,000 acre-feet of water are displaced each year by accumulating sediment.[44]

But the basic problem is not simply about providing sufficient water for Tucson, or any other western city. It is not about improving water consumption habits, developing new technologies, or finding ways to make reservoirs and water transportation systems more efficient. The basic problem is much larger and in fact involves a set of nested problems. What needs to be better addressed are the social and environmental implications of removing water for desert use on all the environments and communities affected by that removal.

By taking a share of the Colorado River's flow, Tucson contributes to any number of environmental problems in other places. The Colorado River Delta in Mexico, for example, where habitat for migratory birds has shrunk, needs to be revitalized by regular and larger flows of Colorado River water and the sediment loads it once contained. In the Sea of Cortez, fish such as the totoaba and a species of porpoise are endangered, in part because of Colorado River management practices. Further north and east, Black Mesa is no longer strip-mined to provide power to pump water to Tucson, but its coal may be put to some other purpose if fossil fuel economies change once again. Throughout the Colorado River system, water diversions have lowered water quality to such an extent that in many places water is too salty to meet the requirements of the Clean Water Act. Such examples could be extended.

Yet as Colorado River water entered Tucson's municipal system for the second time, there was very little local public discussion of the sources and sustainability of CAP water itself, or with the effect that Tucson has on distant environments.

Instead, public discussion invoked past problems—of taste, corrosion of pipes, and damage to appliances—and celebrated efforts to overcome them. With mixed feelings, but also with general support, Tucsonans anticipated the future benefits of CAP water.[45]

In a generation or two, will these future benefits still obtain?

Public Discourse and Environmental Values

Missing in public discourse in the late-twentieth and into the twenty-first century was any sustained debate about the environmental values that informed the construction of the CAP in the first place, or about the basis for the governmental decisions that at great expense moved water to Tucson. This is unfortunate because current residents of Tucson confront those values each time they turn on a water tap, jump into a swimming pool, and irrigate their gardens.

Missing too was sustained discussion of the larger environmental responsibility Tucsonans share with all residents within the Colorado River plumbing system. This is also unfortunate, since Tucsonans are now as culpable for the effects of water storage, diversions, and withdrawals as those in southern California, who take more than their allotted 4.4 million acre-feet each year, or those in Denver who benefit from 17 transmountain water diversions that transport water between different hydrological systems, or those millions of persons who each year recreate on huge bodies of Colorado River water in its placid, domesticated, water-skiing, pleasure-boating form—on Lakes Havasu, Mead, and Powell. Tucsonans share a wider guilt in the environmental effects of their water use, but there is no scapegoat, no single governmental body to blame. All users are implicated in the widespread environmental and social effects of the Colorado plumbing system, yet a collective sense of shared responsibility appears to be absent.

Rivers connect diverse environments. They also connect different polities, cultures, and histories. But they flow, or fail to flow, based on any number of unpredictable natural and human-induced changes to the world. Technological fixes to the unpredictable flows of rivers bring with them their own set of contingencies, as do the demands of increasing human populations. For the Colorado River, the result can best be characterized as a set of competing interests, worked out through compacts, laws, treaties, and in courts—the so-called Law of the River. A set of common social or environmental values, by contrast, does not characterize uses of the river.

One explanation for the split between local interests and a wider set of shared values is historical. Current generations take for granted the plumbing system, unless it fails, and may not clearly understand the large-scale environmental consequences of the choices of their ancestors. Such choices and their consequences do not remain at the forefront of public discussion, despite the considerable efforts of environmental groups to keep them there.

A second explanation for the lack of shared social and environmental values is geographical. Local people tend to stay concerned with local environments and often do not extend their concerns to distant locales unless, again, the system fails. In this, Tucson is no different from other western cities.

A third explanation is ideological. By casting arguments about the Colorado River as natural system vs. a plumbing system, the debate about the future of the river and its many uses becomes polarized and the essential connection between nature and culture is obscured. Even at its most wild, the Colorado River is a now plumbing system. Even in its most domesticated form, the Colorado River is still part of the natural world.

The Worth of Water

People in Tucson will, more likely than not, adopt water conservation practices little by little. They will improve drip irrigation for their gardens, and place rain barrels beneath their waterspouts. Water systems that bring potable water to Tucson's households will be decoupled from water systems for golf courses and parks—a decoupling Tucson has already begun. Perhaps inflatable dams will be built, in effect creating riverine rain barrels, so that rainwater flushed off Tucson's streets will be put to some purpose. Over time, as people conserve water and the population grows, the city may turn brown and dun colored, as water-greedy green lawns, plants, and trees are replaced with drought-resistant species, some of which are local, some of which hail from other parts of the planet. Rising energy cost may become the strongest water conservation factor, forcing people to deal more effectively with the essential aridity of the southwest simply because they cannot afford to do otherwise.

Major Powell, hero of western river runners who frequently cite his account of the Green and Colorado Rivers, would not have been surprised at the fate of the Colorado River. "All the waters of all the arid lands," he predicted in his report to Congress, "will eventually be taken from their natural channels" and used for agricultural and other human purposes. He might have been surprised, however, at the means to that fate. Powell believed the future of western development should not be in the hands of the federal government, but in private hands. Under his proposal, large rivers such as the Colorado may have become mere rivulets, when all their tributaries were blocked with relatively small dams for local use. The opposite of Powell's vision is the contemporary reality. The results of western water policies are massive dams such as Hoover, Glen Canyon, Parker, and Flaming Gorge, massive water projects such as the Central Arizona Project, the Central Utah Project, and others, coordinated by federal agencies and funded by federal dollars, and the apparently unsustainable growth of cities such as Tucson, Phoenix, and Las Vegas.

It is not Powell and his vision for the arid West but rather an earlier American, Benjamin Franklin, who best characterized the difficulty with sustainability in a place of little rain, and whose social and political trajectories appear to ignore that central fact. "When the well's dry," he said in one of his famous aphorisms, "we know the worth of water."

Notes

1 Despite human encroachment, the Sonoran Desert is still extraordinarily diverse, with a large variety of flowering trees and plants; hundreds of species of bees, butterflies, and moths; dozens of reptile species; some 86 species of mammals; and approximately 450 nesting and migrating bird species. See Steven J. Phillips and Patricia Wentworth Comus, eds., *A Natural History of the Sonora Desert* (Tucson, AZ: Arizona-Sonora Desert Museum Press, 2000). See also Gary Paul Nabhan's engaging book *The Desert Smells Like Rain: A Naturalist in O'Odham Country* (San Francisco, CA: North Point Press, 1982).

2 On Father Kino, see Herbert E. Bolton, *Rim of Christendom: A Biography of Eusibio Francisco Kino, Pacific Coast Pioneer* (New York, NY: Macmillan, 1936; reprint, Tucson, AZ: University of Arizona Press, 1984); On the Hohokam, see Jefferson Reid and Stephanie Whittlesey, *The Archaeology of Ancient Arizona* (Tucson, AZ: University of Arizona Press, 1997). On the establishment of non-native species, see Phillips and Comus, eds., *A Natural History of the Sonoran Desert*.

3 Participating groups in the international project included the Center for the Study of Developing Societies in India, the Lake Biwa Museum in Japan, the Research Center for Contemporary China, at Peking University, and the Bureau of Applied Research in Anthropology, at The University of Arizona. We wanted to know how local values entered the policymaking process, whether environmental concerns of local people were being adequately addressed by policymakers, and whether there were any commonalities across ten markedly different study sites in these four countries. See the project description at the website for the Carnegie Council on Ethics and International Affairs, www.cceia.org, and the volume that resulted, Joanne Bauer, ed., *Forging Environmentalism: Justice, Livelihood, and Contested Environments* (New York, NY: M.E. Sharp, 2006). For an extended analysis of sustainability in Tucson, see in the same volume, David Jenkins, Joanne Bauer, Scott Brunton, Diane Austin, and Thomas McGuire, "Two Faces of American Environmentalism: The Quest for Justice in Southern Louisiana and Sustainability in the Sonoran Desert." See also David Jenkins, "Atlantic Salmon, Endangered Species, and the Failure of Environmental Policies," *Comparative Studies in Society and History* 45: 843–872 (October 2003), a study which originated in the larger project.

4 For a recent assessment, see Colin Strong, Samantha Kuzman, Samuel Vionnet, and Paul Reig, "Achieving Abundance: Understanding the Cost of a Sustainable Water Future," *World Resources Institute Working Paper* (January 2020).

5 For water use figures, see "Water in the West: Challenge for the Next Century: Report of the Western Policy Review Advisory Commission" (June 1998). On Tucson water consumption, see William E. Martin, Helen M. Ingram, Nancy K. Laney, and Adrian H. Griffin, *Saving Water in a Desert City* (Washington, DC: Resources for the Future, 1984). The latest figures, compiled by the U.S. Geological Survey, are available at https://map azdashboard.arizona.edu/infrastructure/residential-water-use.

6 On the history of water use in the area, see Joe Gelt, Jim Henderson, Kenneth Seasholes, Barbara Tellman, and Gary Woodard, with Kyle Carpenter, Chris Hudson, and Souad Sherif, "Water in the Tucson Area: Seeking Sustainability," *Water Resources Research Center Issue Paper No. 20* (1999); Joe Gelt, "Water Conservation, Yesterday and Today: A Story of History, Culture and Politics," *Arroyo* 10 (December 1999); T. Lindsay Baker, Steven R. Rae, Joseph E. Minor, and Seymour V. Connor, *Water for the Southwest: Historical Survey and Guide to Historic Sites* (New York, NY: American Society of Civil Engineers, 1973).

7 On the Ogallala aquifer, see John Opie, *Ogallala: Water for a Dry Land* (Lincoln, NE: University of Nebraska Press, 1993). The 1983 data for the Tucson area are from William E. Martin, Helen M. Ingram, Dennis, C. Cory, and Mary G. Wallace, "Toward Sustaining a Desert Metropolis: Water and Land Use in Tucson, Arizona," in *Water and Arid Lands of the Western United States*, Mohamed T. El-Ashry and Diana C. Gibbons, eds. (Cambridge: Cambridge University Press, 1988), 281–327. In 1983, the total cost (pumpage, administrative, distribution, capital repayment) to deliver water was approximately $1,334.93 for a million gallons. Liz Greene, of Tucson Water, provided the 2000 pumpage data. On the projected electrical costs for Tucson Water, see David Modeer, "Power, New Sources Boost Water Costs," *Arizona Daily Star* (February 23, 2001), B7. See also Martin, Ingram, Laney, and Griffin, *Saving Water in a Desert City*.

8 R.T. Hanson and J.F. Benedict, *Simulation of Ground Water Flow and Potential Land Subsidence, Upper Santa Cruz Basin, Arizona* (Tucson, AZ: U.S. Department of Interior, U.S. Geological Survey, 1994).

9 See *Water Transfers in the West: Efficiency, Equity, and the Environment*, (Washington, DC: National Academy Press, 1992), especially Chapter 9, "Central Arizona: The Endless Search for New Supplies to Water the Desert." See also *Third Management Plan for Tucson Active Management Area, 2000-2010* (Arizona Department of Water Resources, December 1999); *Third Management Plan for Pinal Active Management Area, 2000-2010* (Arizona Department of Water Resources, December 1999); *Third Management Plan for Phoenix Active Management Area, 2000-2010* (Arizona Department of Water Resources, December 1999); *Third Management Plan for Santa Cruz Active Management Area, 2000-2010* (Arizona Department of Water Resources, December 1999); *Third Management Plan for Prescott Active Management Area, 2000-2010* (Arizona Department of Water Resources, December 1999).

10 Martin, Ingram, Cory, and Wallace, "Toward Sustaining a Desert Metropolis," 311–312.

11 Andrew C. Comrie, "Mapping a Wind-Modified Urban Heat Island in Tucson, Arizona (with Comments on Integrating Research and Undergraduate Learning)," *Bulletin of the American Meteorological Society* 81:1–15 (October 2000).

12 *Third Management Plan for Pinal Active Management Area, 2000-2010*, 1–1, 1–2.

13 On the history of Colorado River water, see Philip Fradkin, *A River No More: the Colorado River and the West* (New York, NY: Knopf, 1981). On the current state of the river see Dale Pontius, with SWCA, Inc., "Colorado River Basin Study, Report to the Western Water Policy Review Advisory Committee," (August 1997). Joe Gelt provides an accessible summary, "Sharing Colorado River Water: History, Public Policy and the Colorado River Compact," *Arroyo* 10 (August 1997). See Franck Poupeau, et al., eds., *Water Bankruptcy in the Land of Plenty* (New York, NY: Routledge, 2017) for a recent assessment. April R. Summitt also provides an overview, *Contested Waters: An Environmental History of the Colorado River* (Boulder, CO: University Press of Colorado, 2013). On restoration, see Robert W. Adler, *Restoring Colorado River Ecosystems: A Troubled Sense of Immensity* (Washington, DC: Island Press, 2007). See also David H. Getches and Charles J. Meyers, "The River of Controversy: Persistent Issues," and Norris Hundley, Jr., "The West against Itself: The Colorado River—An Institutional History," both in *New Courses for the Colorado River: Major Issues for the Next Century*, Gary D. Weatherford and F. Lee Brown, eds. (Albuquerque, NM: University of New Mexico Press, 1986). The *Tucson Weekly* consistently ridiculed CAP water initiatives. See "Vote Yes on Prop 200" (November 2–8, 1995), Jim Wright, "Pumping Money" (May 2–8, 1996), Vicki Hart, "CAP Is Still Crap" (October 23–29, 1997), Vicki Hart, "Pumping Bile," and Jim Nintzel, "Flow Chart" (August 19–25, 1999).

14 *Colorado River Compact*, 1922, 45 Stat. 571. The text of the compact is available at www. lc.usbr.gov. See Norris Hundley, Jr., *Water and the West: the Colorado River Compact and the Politics of Water in the American West* (Berkeley, CA: University of California Press, 1975).

15 Donald Worster, *A River Running West: The Life of John Wesley Powell* (New York, NY: Oxford University Press, 2001), 249. See also Juanita Brooks, *The Mountain Meadow Massacre* (Palo Alto, CA: University of California Press, 1950), and *John Doyle Lee: Zealot, Pioneer, Builder, Scapegoat* (Glendale, CA: A.H. Clark, 1962).

16 *Boulder Canyon Project Act*, 45 Stat. 1057, 43 USC 617. California and Arizona had a long-running dispute over appropriate percentages of water, which was finally resolved in 1963 by the Supreme Court in *Arizona v. California*, 373 U.S. 546 (March 9, 1964). Both the act and the judicial decision can be found at www.lc.usbr.gov. On the history of the Mexican treaty, see Norris Hundley, Jr., *Dividing the Waters: A Century of Controversy between the United States and Mexico* (Berkeley, CA: University of California Press, 1966).

17 For a study of long-term flow measurements, see David Meko, Charles W. Stockton, and William R. Burgess, "The Tree-Ring Record of Severe Sustained Drought," *Water Resources Bulletin* 31: 789–801 (1995), and David M. Meko, et al., "Medieval Drought in the Upper Colorado River Basin," *Geophysical Research Letters* 34:10 (2007). See generally *Water in the West: Challenge for the Next Century*. See also Donald Worster, *Rivers of Empire: Water, Aridity, and the Growth of the American West* (New York, NY: Oxford University Press, 1985).

18 For histories of CAP water see Ernest A. Engelbert, *The Origins and Policy Issues of the Pacific Southwest Water Plan* (Boulder, CO: University of Colorado Press, 1965); Rich Johnson, *The Central Arizona Project, 1918-1968* (Tucson, AZ: University of Arizona Press, 1977); Robert Dean, "'Dam Building Still Had Some Magic Then': Stewart Udall, the Central Arizona Project, and the Evolution of the Pacific Southwest Water Plan, 1963–1968," *Pacific Historical Review* 66:81–98 (February 1997).

19 James Lawrence Powell provides the best history of Lake Powell, *Dead Pool: Lake Powell, Global Warming, and the Future of the West* (Berkeley, CA: University of California Press, 2008).

20 *Colorado River Basin Project Act*, Public Law 90–537, 82 Stat. 885. As Byron E. Pearson points out, the political circumstances were more complex than a simple story of triumphant environmental groups rallying public support to stop the proposed dams; see *Still the Wild River Runs: Congress, the Sierra Club, and the Fight to Save Grand Canyon* (Tucson, AZ: University of Arizona Press, 2002). See also Wendy Nelson Espeland, *The Struggle for Water: Politics, Rationality, and Identity in the American Southwest* (Chicago, IL: University of Chicago Press, 1998).

21 On Indian water rights, see Daniel McCool, *Command of the Waters: Iron Triangles, the Federal Water Development Program, and Indian Water* (Berkeley, CA: University of California Press, 1987), Thomas R. McGuire, William B. Lord, and Mary G. Wallace, eds., *Indian Water in the New West* (Tucson, AZ: University of Arizona Press, 1993), Monroe B. Price and Gary D. Weatherford, "Indian Water Rights in Theory and Practice: Navajo Experience in the Colorado River Basin," *Law and Contemporary Problems* 40:108–131 (1976). On the Winters Doctrine see Norris Hundley, Jr., "The 'Winters' Decision and Indian Water Rights: A Mystery Reexamined," *Western Historical Quarterly* 13:17–42 (1982), and John Shurts, *Indian Reserved Water Rights: The Winters Doctrine in Its Social and Legal Context, 1800s–1930s* (Norman, OK: University of Oklahoma Press, 2000).

22 Some people argue that it is highly improbable that California would insists on receiving all its 4.4 million acre-feet at the expense of Arizona during a prolonged drought. See "The Colorado River Compact at 75: A Conversation About Its Past and Future,"

7–8. Convened by the Western Water Policy Commission of the Council of State Governments—WEST (August 22, 1997).

23 Fradkin, *A River No More*, 3–14. Espeland, *The Struggle for Water*, 4–14.

24 The Bureau of Reclamation initially owned 24.3 percent of the Navajo Generating Station; the Salt River Project owned 21.7 percent; the Los Angeles Department of Water and Power owned 21.2 percent; the Arizona Public Service Company owned 14 percent; Nevada Power Co. owned 11.3 percent; and Tucson Gas and Electric owned 7.5 percent. The Navajo Generating Station was decommissioned in 2019. For an analysis of the Ninth Circuit's decision concerning air pollution from the Navajo Generating Station, in *Central Arizona Water Conservation District v. EPA*, see R. Nicole Cordan, "Lost in the Haze? Central Arizona Fulfills Congress's Promise to Protect Visibility in the National Parks," *Environmental Law* 24:1371–1394 (July 1994).

25 Richard White, *"It's Your Misfortune and None of My Own": A History of the American West* (Norman, OK: University of Oklahoma Press, 1991), 558. See also Susanne Gordon, photographs by Alan Copeland, *Black Mesa: Angel of Death* (New York, NY: Double Day, 1973). Indians continued to lose. Peabody Coal pumped 4,000 acre-feet of groundwater each year to put in its pipeline. This is pristine water from the N-aquifer that locals use for drinking. See David Beckman, Michael Jasny, Lissa Wadewitz, and Andrew Wetzler, "Drawdown: Groundwater Mining on Black Mesa" (Natural Resources Defense Council, October 2000).

26 Ryan Randazzo, "Water Users Better Off Without Navajo Plant," *The Republic* (February 16, 2017).

27 Ryan Randazzo and Shondiin Silversmith, "Navajo Generating Station—The Largest Coal Plant in the West—Has Shut Down." *The Republic* (November 18, 2019).

28 Marc Reisner, *Cadillac Desert: The American West and Its Disappearing Water* (New York, NY: Viking, 1986), 304.

29 Maurice M. Kelso, William E. Martin, and Lawrence E. Mack, *Water Supplies and Economic Growth in an Arid Environment: An Arizona Case Study* (Tucson, AZ: University of Arizona Press, 1973). See the series on CAP water in the *Arizona Daily Star* (April 29–May 4, 2001). See also J.L. Barr and D.E. Pingry, "The Central Arizona Project: An Inquiry into Its Potential Impacts," *Arizona Review* (1977).

30 See various articles by Mary Manning, "Suits Filed against Radioactive Flows into Colorado River," *Las Vegas Sun* (October 23, 1998), "Officials Fear Floods Could Cause Radioactive Contamination of Water," *Las Vegas Sun* (July 30, 1999), "So. California Backs Bill on Radioactive Water Tailings," *Las Vegas Sun* (February 10, 1999). The law requiring the Department of Energy to clean up the tailings was part of the Floyd D. Spence National Defense Authorization Act for Fiscal Year 2001, Public Law: 106-398. Energy Secretary Bill Richardson, in a Department of Energy press release, seemed to give more weight to scenic protection than to water purity:

> Radioactive waste sits at the gateway of two national parks, Arches and Canyonlands. This area is a geological wonderland, nestled in a valley with scenic red cliffs and surrounded by rugged, beautiful desert terrain. The Department of Energy has the expertise and experience to relocate the material in a secure, permanent location that is safely away from the Colorado River and the national parks.
>
> *www.energy.gov/HQPress/releasses00/janpr/pr00009.htm*

See the Department of Energy website devoted to the Moab tailings, www.gjo.doe. gov/moab, and The National Academy of Sciences June, 2002 report, "Action at the Moab Site—Now and for the Long Term" (The National Academies, Committee on Long-Term Institutional Management of DOE Legacy Waste Sites: Phase 2).

31 *Water Consumer Protection Act*, Public Initiative Petition 1994–2001.

32 For an interesting and helpful assessment of sustainable water, see Jason I. Morrison, Sandra L. Postel, and Peter H. Gleick, *The Sustainable Use of Water in the Lower Colorado River Basin* (Oakland, CA: Joint Report of the Pacific Institute for Studies in Development, Environment, and Security and the Global Water Project, November 1966). On the inflatable dam proposal, see "Summary, Rillito Recharge Project: Artificial Groundwater Recharge Demonstration Project," and "Final Report: Rillito Recharge Project," U.S. Department of the Interior, Bureau of Reclamation, in participation with the U.S. Environmental Protection Agency, November 1996.

33 For hopeful newspaper accounts, see Maureen O'Connell, "Blended Water to Flow Here by Spring," *Arizona Daily Star* (September 25, 2000), and Mitch Tobin, "CAP Water to be Clean, Leaders Vow," *Arizona Daily Star* (March 2, 2001). See also the series on CAP water in the *Arizona Daily Star* (April 29–May 4, 2001). The City of Tucson CAP website has up-to-date use information. www.tucsonaz.gov/water/recharged-water.

34 For the most recent data on groundwater replenishment, see www.tucsonaz.gov/water/groundwater-recovery. See A. Park Williams, et al., "Large Contribution from Anthropogenic Warming to an Emerging North American Megadrought," *Science* 368 (April 17, 2020). See also the Colorado River Water Users Association website, www.crwua.org/.

35 On the uncertainty of sustained Colorado River flows, see Tim P. Barnett and David W. Pierce, "When Will Lake Mead Go Dry?" *Water Resources Research* 44:3 (March 28, 2008). On the effects of mixing CAP water and groundwater, see David Devine and Molly Mckasson, "What Does the Future Hold for Tucson's Water Supply? Only One Thing Is Certain: It's All Going to Cost More," *Tucson Weekly* (March 6, 2008), and "Will Tucson Water Spend More to Increase the Water Supply—Or Will It Spend More to Improve the Product?" *Tucson Weekly* (March 24, 2011).

36 Eric Kuhn and John Fleck, in *Science be Dammed: How Ignoring Inconvenient Science Drained the Colorado River* (Tucson, AZ: University of Arizona Press, 2019), 28.

37 Kuhn and Fleck, *Science be Dammed*, 43.

38 Kuhn and Fleck, *Science be Dammed*, 84, 132 ff.

39 Morris K. Udall, Bob Neuman, and Randy Udall, *Too Funny to be President* (New York, NY: Henry Holt and Company, 1988), 61.

40 The literature on the topic is expanding rapidly. See Daniel B. Botkin, *Discordant Harmonies: A New Ecology for the Twenty-First Century* (New York, NY: Oxford University Press, 1990). For studies of human impacts on local ecosystems see Carole L. Crumbly, ed., *Historical Ecology: Cultural Knowledge and Changing Landscapes* (Santa Fe, NM: School of American Research, 1994); Jeanne X. Kasperson, Roger E. Kasperson, and B.L. Turner, eds., *Regions at Risk: Comparisons of Threatened Environments* (Tokyo, New York, NY, Paris: United Nations University Press, 1995); Shepard Krech III, *The Ecological Indian: Myth and History* (New York, NY: W.W. Norton, 1999); Mark J. McDonnell and Steward T.A. Pickett, eds., *Humans as Components of Ecosystems: The Ecology of Subtle Human Effects and Populated Areas* (New York, NY: Springer-Verlag, 1993); J.R. McNeill, *Something New Under the Sun: An Environmental History of the Twentieth-Century World* (New York, NY: W.W. Norton, 2000); Charles L. Redman, *Human Impacts on Ancient Environments* (Tucson, AZ: The University of Arizona Press, 1999); B.L. Turner II, William C. Clark, Robert W. Kates, John F. Richards, Jessica T. Mathews, and William B. Meyers, eds., *The Earth as Transformed by Human Action: Global and Regional Changes in the Biosphere over the Past 300 Years* (Cambridge: Cambridge University Press, 1990). Peter M. Vitousek, Harold A. Mooney, Jane Lubchenco, and Jerry M. Melillo, "Human Dominations of Earth's Ecosystems," *Science* 227:494–499 (1997).

41 John Wesley Powell, "Report on the Lands of the Arid Region," 45 Cong., 2nd session, House Exec. Doc. 73. Worster, *A River Running West*, 354–360.

42 For an example of a subtle and apparently sustainable system of allocating water, see J. Stephen Lansing, *Priests and Programmers: Technology of Power in the Engineered Landscape of Bali* (Princeton, NJ: Princeton University Press, 1991).

43 On artificial floods, see W.K. Stevens, "Grand Canyon Roars Again as Ecologic Clock Is Turned Back," *The New York Times* (February 25, 1997); on introduced and native fish, see W.L. Minckley, "Native Fishes of the Grand Canyon: An Obituary?" in *Colorado River Ecology and Dam Management* (Washington, DC: National Academy Press, 1977), 124–177; Richard White, *The Organic Machine: Making and Remaking the Columbia River* (Hill and Wang, 1996).

44 On sediment flows see Fradkin, *A River No More*, 182.

45 For a discussion of the Colorado River Delta, see Jennifer Pitt, Daniel F. Luecke, Michael J. Cohen, Edward P. Glenn, and Carlos Valdés-Casillas, "Two Nations, One River: Managing Ecosystem Conservation in the Colorado River Delta," *Natural Resources Journal* 40:819–864 (Fall 2000); and Edward P. Glenn, Christopher Lee, Richard Felger, and Scott Zengel, "Effects of Water Management on the Wetlands of the Colorado River Delta, Mexico," *Conservation Biology* 10:1175–1186 (August 1996).

3
ATLANTIC SALMON, ENDANGERED SPECIES, AND THE FAILURE OF ENVIRONMENTAL POLICIES

The struggle to control nature takes many forms. On the east coast of North America, a decades-long struggle to control nature swirled around declining stocks of Atlantic salmon. In the State of Maine, the varied participants in the debates and struggles over Atlantic salmon were myriad and included federal and state governments and many of their bureaucracies, large international timber corporations, small local businesses, salmon farmers, dam owners, blueberry growers, commercial fishers, recreational fishers, scientists, and a raft of environmental organizations.

In different ways, these participants all had a stake in the fate of Atlantic salmon. They did not, however, have the same power to effect ecological change or to define the debate in terms most favorable to their wishes.[1]

Contested definitions of the natural world and its uses characterize much of the discourse around land management bureaucracies. This is one reason I describe traditional bureaucratic knowledge as residing between knowledge systems, especially between economic, scientific, political, and cultural knowledge systems. There is a constant interplay among these domains of knowledge and action. Sometimes the result is agreement, sometimes protracted contestation. Land management bureaucracies are often at the center of both.

For Atlantic salmon, arguments were common over definitions of "wild" vs. "farmed" salmon, "native" salmon, "aboriginal stock," "species," and "endangered," "threatened," and "extinct" populations. The disputes, however, expanded beyond semantics. They were not solely scientific. At stake was who or what governmental entity had the power to force its definitions on everyone else. The definitions matter. Environmental consequences flow from them.

A modern scientific assessment of Atlantic salmon directly informed bureaucratic decisions about the fish. Diverse stakeholders also actively made use of that same science for their own purposes. The result was a contested endangered species listing of distinct populations of Atlantic salmon, with each side asserting

DOI: 10.4324/9781003297444-5

the scientific high ground. Despite the arguments, certain populations of Atlantic salmon were determined to be endangered in November 2000. They have remained in that status ever since.

The Endangered Species Act, by requiring species and habitat restoration to be cast in terms of the best science of the day, polarizes scientific discourse for political ends. The listing of Atlantic salmon provides one example of such polarization. Many participants argued about science, when their concerns and motivations were not scientific. Problems of scientific terminology merged with debates over proper governance and power, and with the imposition of limits to human behavior.

This chapter traces part of the history of salmon population declines, with the aim of contextualizing contemporary debates about the endangered species listing. These debates reveal that much of the recent controversy over Atlantic salmon had at its core a debate about proper governance. They demonstrate that to control nature is simultaneously to control human beings. Finally, this chapter shows that endangered species listings are the result of past failures and poor decisions and are themselves symbolic of those failures and those decisions.

First, I sketch out the terms of contemporary argument.

What Is a Wild Salmon?

For some researchers and environmental advocates, *wild* salmon were those that had not been substantially affected by human activities. They lived their lives—from natal stream to ocean and back to their natal stream to spawn—outside of human influence. They maintained a genetic link to similar, wild ancestors. They were affected by pressures of natural selection and thus exhibit genetic robustness. They had river-specific characteristics that differentiated them from other populations of Atlantic salmon.

Under this definition, a mere 100 wild salmon were estimated to have returned to seven Maine rivers to spawn in 2000. Wild females produce about 7,200 eggs. If half of the returning wild salmon in 2000 were female, then 360,000 eggs were potentially deposited. Assuming each egg was fertilized, the result was a substantial number of young salmon, called fry at the initial free-swimming stage of their complex lives. The best estimate, however, is that 50,000 fry result in one returning adult fish. Under ideal circumstances, 360,000 fry yielded seven adults which might return in four, five, or even six years to spawn. This rate of return was by no means sustainable.[2]

Yet definitions of wild salmon need not be so restrictive. A wild salmon may be one whose parents lived a natural life cycle characteristic of anadromous fish: natal stream to ocean and back again to spawn. Under this definition, it does not matter if there was an ancestral link to aboriginal fish. Wild salmon could equally come from stocked fish, or fish escaped from salmon farms.

Over the last century, more than 100 million salmon, many with origins in Canadian rivers, had been stocked in Maine rivers, with strikingly poor but measurable results. In addition, an apparently small number of salmon that escaped from salmon farms along Maine's coast migrated to Maine rivers—fewer than 100 escaped farmed salmon were documented in two Maine rivers between 1993 and 1997. The potential for large numbers of farm-raised salmon to enter Maine rivers remained high, however. More than 100,000 farmed salmon escaped from pens in Cobscook Bay near the Canadian border during a severe winter storm in 2000, some of which may have migrated to Maine rivers. Under the less restrictive definition, any descendants of stocked or escaped fish would be wild salmon.[3]

The first definition of wildness relies on genetic continuity with an ancestral population, the second is behavioral. Each makes sense. But different social and environmental consequences result from them, depending upon which definition becomes codified into policy. It is here that governmental bureaucracies are key players.

Through the U.S. Fish and Wildlife Service and the National Marine Fisheries Service (the Services), the federal government advanced a definition based on genetic continuity, which served as one rationale for listing Atlantic salmon as endangered under the Endangered Species Act. After considerable scientific review, a Gulf of Maine population of Atlantic salmon was listed as endangered in November 2000, reversing a 1997 determination that a listing was not justified.

The State of Maine rejected the federal government's 2000 determination, as well as the July 1999 status review and the November 1999 proposed rule on which it was based and argued that there was no genetic continuity from aboriginal to contemporary salmon stocks, or at least that it could not be conclusively demonstrated. Through state and federal hatchery programs, Maine argued, humans had been manipulating salmon genetics for over a century, and there was no scientific basis for claiming that any salmon returning to Maine rivers fit the appropriate definitions under the Endangered Species Act. Therefore, the Act could not be used as a legally appropriate mechanism for restoring salmon populations.[4]

Both state and federal officials recognized that the long-term survival of salmon was doubtful. Over the last century, state efforts to restore Atlantic salmon populations to New England rivers had failed. Over the same period of time, federal hatchery efforts to restore salmon populations had failed. Industry attempts to mitigate damage to salmon habitats had likewise failed. Wild Atlantic salmon had nearly disappeared from U.S. rivers.

Disappearing Salmon

However defined, wild Atlantic salmon were clearly on the decline. Historically, salmon were native to most major river systems north of the Hudson River. By 1865, salmon had vanished from rivers in southern New England, largely because of fishing, pollution, and dams. By 1900, wild salmon no longer spawned in the Connecticut, Merrimack, or Androscoggin Rivers. Today, their range in the United States has shrunk to a few rivers in northeastern Maine. Between 28 and 34 Maine rivers used to have wild salmon populations. By the 1870s, only seven or eight contained small, remnant populations of native Atlantic salmon that have, in small numbers, continued to spawn to the present day.[5]

Responses to declining stocks of Atlantic salmon have a long history and include various local regulations requiring adequate passages for fish around man-made obstacles such as dams and governing net sizes. The town of Machias, Maine, for example, enacted in 1780 a regulation to ensure that fish could migrate past dams on the Machias River. After initial success, the regulation proved ineffective. Two new dams were built in 1841–1842, and salmon virtually disappeared from the Machias River. They returned in diminished runs, only when a fishway was constructed in the 1870s.

In 1814, nets in the Penobscot River were limited by statute "to one-third of the width of the stream where used," as Charles G. Atkins mentioned in his nineteenth century assessment of the fisheries of Maine. Salmon nevertheless declined sharply in the river.[6] Two hundred thousand pounds of salmon were caught in the Penobscot River in 1888. Ten years later, only 53,000 pounds were harvested. By 1948, the catch was down to 400 pounds, or about 40 fish.

In his 1872 review of the status of salmon in Maine, Atkins listed the following rivers to which "salmon are now regular visitors": the Saint Johns, Saint Croix, Denny's, Little Falls, East Machias, Wescongus, Penobscot, and Kennebec. Atkins noted that other rivers had occasional salmon visitors, but

> the ancient brood of salmon was long ago extinguished, and the rare specimen occasionally observed must be regarded either as strays from some of the better-preserved rivers, or as early-returning members of the new broods established by artificial culture in several rivers.

There was, Atkins thought, some reason for hope. With new scientific understanding of anadromous fish, improved passage for migration, and cooperation between state and federal agencies, runs of Atlantic salmon could be restored—if not to former levels, at least to levels that would be self-sustaining with a little human intervention. Atkins spent his adult life trying to restore salmon and many other species of fish to prior levels of abundance. He died in 1921, having achieved over a 50-year career modest success in restoring salmon runs.[7]

Atkins attributed salmon declines to a number of causes, including overfishing, timber harvesting, dams, and pollution from manufacturing. Timber harvesting and sawmills in nineteenth-century Maine were responsible for dramatically altering river ecosystems. Anthony Netboy, in his world-wide study of Atlantic salmon, writes that Maine

> rivers were slowly accumulating mountains of sawdust and other lumbering debris, creating sandbars and narrowing the channels, making fish migration difficult. In 1834 a state law was enacted prohibiting the practice of dumping [lumbering debris] on the Kennebec [River], and similar laws came later to protect other streams, but it is doubtful if they were enforced or did much good to abate pollution.[8]

The massive accumulation of lumbering debris was not the only problem for Atlantic salmon. Unlike the Great Lakes region and the Pacific Northwest, where lumbermen adopted railroads as the most efficient means to transport logs to mills, Maine lumbermen relied on river transportation. Rivers determined the pattern of logging but needed to be widened, dammed, and cleared of rock obstructions so that logs could be more easily floated downriver and supplies ferried upriver. Log drives effectively scoured-out rivers, changing river contours and sediment patterns, fouling the water, and severely diminishing riverine habitat. One lumberman of the time described the power of a log drive. "It is great," he recalled,

> to see a great jam of logs moved by a shot from a dam—that is, by opening the gates and letting the water out. It is marvelous the force the water has got when it is turned loose. You will see big logs snap and break in two; and you will see big trees standing on the bank of a stream, and logs will run in behind them, and those trees will flatten down and be torn out by the roots.[9]

Timber harvesting and river transportation of logs altered the physical characteristics of rivers and riverbeds, with unknown consequences for salmon. Dams, by contrast, presented obvious obstacles to salmon migration. Netboy, citing E.M. Stillwell's 1872 survey of salmon rivers, noted that nineteenth-century Maine was "forested with mill dams." These dams, most of which were too high for salmon migration, provided power for tanneries, ironworks, paper mills, grist mills, cotton mills, woolen mills, and sawmills. With few exceptions, the economic needs of a developing state superseded the ecological needs of anadromous fish. Timber resources, and dams and the power they provided, were more valuable than fish and more important than the health of a river. Still, there was widespread recognition by the late-nineteenth century that a problem existed, that salmon were disappearing, and that humans were to blame.[10]

The most significant early response to declining stocks was the development of federal and state hatchery programs and fisheries commissions. In the 1860s, New Hampshire, Massachusetts, Vermont, and Connecticut appointed commissioners to study ways to restore migratory fishes to state rivers. Congress created the U.S. Fish Commission in 1870, which would study problems associated with declining runs and find ways to facilitate cooperation between states to improve passage for fish.

The task of the Fish Commission soon became, however, to make and distribute fish, as historian Joseph Taylor has noted in his book *Making Salmon*. In the late-nineteenth century, making fish was seen as the solution to depleted fisheries. Science and technology would rise to the task and provide appropriate techniques to hatch and release fish, thereby augmenting and improving stocks. The environmental effects of overfishing, dams, and industry could be overcome by an emerging science of fish culture. Rather than carefully and explicitly regulating human behavior and the practices that resulted in declining fisheries, the more significant task became the manipulation of salmon.

In 1857, the first anadromous fish to be hatched was Atlantic salmon. By 1868, salmon eggs from New Brunswick, Canada, were being distributed to New England states. In 1871, Maine, with support from Massachusetts and Connecticut and under the direction of Charles Atkins, constructed a hatchery at Craig Brook Pond, in recognition that self-perpetuating salmon were in serious trouble. Initial efforts were of limited success. The next year, the operations were moved to Bucksport, Maine, with additional support from Rhode Island and the U.S. Commission of Fish and Fisheries.[11]

Nineteenth-century hatchery techniques were technologically simple. They required the removal of eggs and milt from adult salmon, the construction of environments in which fertilized eggs would mature, and the means to ship delicate eggs to rivers in need of restoration. Atkins' vivid 1872 description is worth reproducing:

> The mode of manipulation adopted, as, under all considerations, the best, was the following. The spawn-taker sits on a stool of convenient height, with a shallow ten-quart pan before him. He is so clad that he need not avoid close contact with the wet fish, and when a female salmon is brought him he seizes the tail with his right hand, puts her head under his left arm, and holds the vent over the pan. His left hand is free to press the abdomen and force out the spawn. In this way one man can do the whole work alone, and quite as rapidly as he could with two assistants to hold the fish. The eggs are accompanied by a sufficient quantity of transparent, viscous liquid to insure easy motion in the mass without friction, and to prevent rapid evaporation when they are exposed to the air. The time required to take all the eggs from a single fish varies from five to twenty minutes, depending in a great degree upon the size and disposition of the fish. Sometimes she is exceedingly restive, and in such cases it is found best to suspend pressure while she struggles.[12]

Male fish were subject to a similar procedure. "The spawn-taker seizes a male salmon, holds him over the pan in the same position as the other, and presses out his milt upon the eggs." Males were more difficult to handle, but less time was required to relieve them of their milt. "The males are stronger, and struggle more than the females, but this part of the operation is soon concluded." The eggs and milt were stirred together in the pan, after which water was added, which the fertilized eggs absorbed. The eggs were then poured into pails and stored in the hatching house. In this manner, over a seven-day period in October and November 1872, "nearly a million eggs had been taken."[13]

At the Bucksport facility, eggs continued to be stripped from salmon, and by early February 1873 a total of 1,291,800 were ready for shipment. Packed in moss and sawdust, and double-boxed, the eggs followed nineteenth-century lines of transportation. A portion of the eggs remained at Bucksport, with the balance destined for distant rivers, "152,000 going on a sled to Bangor and thence by rail to Dixfield, Me.; the rest all going from Bucksport to Boston by steamer, and from that point by rail to their several destinations." These destinations included New Hampshire, Vermont, Massachusetts, Rhode Island, Connecticut, New York, New Jersey, Pennsylvania, Ohio, Michigan, and Wisconsin. The government-sponsored transfer of salmon eggs to distant river systems had begun.[14]

In the early 1870s, Congress made an initial appropriation of $15,000 to make and distribute fish. The idea was to bring Pacific salmon eggs to New England and to plant east coast shad eggs in streams that flowed into the Pacific, a kind of reciprocal egg exchange between distant ecosystems. Salmon eggs from Maine were shipped out, eggs from other ecosystems were shipped in. Proportions fluctuated year by year. Incoming eggs came from rivers and hatcheries in Canada, California, Oregon, Washington, and Alaska, and from the Rhine River in Germany. The federal government oversaw the egg exchange as states lacked sufficient resources for the project. They were, however, responsible for enacting local legislation to regulate sport and commercial fisheries and other activities that affected fish populations.[15]

For nineteenth-century fish culturists, it was crucial to find ways to introduce fish into rivers where they might thrive, both to restore fish populations in depleted rivers and to provide sources of food for a growing human population. Chinese success at raising fish was frequently cited as an example to be emulated. In 1876, the U.S. Commissioner of Fish and Fisheries, Spencer F. Baird, identified seven species of fish that required distribution and management, in part to enhance food security:

> These species to which special attention has been directed are the shad (*Alosa sapidissima*), fresh-water herring or alewive, (*Pomolobus pseudoharengus*), striped bass or rock-fish (*Roccus lineatus*), California salmon (*Salmo quinnat*), the salmon of Maine (*Salmo salar*), land-locked salmon (*Salmo sebago*), white-fish (*Coregonus albus*), and the carp (*Cyprinus carpio* and var.), each of these having special relations to certain portions of the country, and promising in their anticipated aggregate an extremely important addition to the food-resources of the United States.[16]

The transportation of these and other fish was subject to the vicissitudes of nineteenth-century modes of travel. Fish-crates, and sufficient ice to keep eggs cool over long journeys, often weighed several tons. In one instance, Livingston Stone, who managed the hatchery on the Sacramento River in California, was traveling west in 1873 with a specially fitted railroad car filled with tanks containing 60 black bass, 11 pike, 190 yellow perch, 12 bullheads, 110 catfish, 20 tautogs, 41,500 eels, 1,000 trout, 20,000 shad, 162 lobsters, and one barrel of oysters. Along the way, over the Elkhorn River in Nebraska, a bridge trestle broke and "the aquarium-car was precipitated into the river, the car was partially up-ended, and the tanks thrown into confusion." The fish escaped. Commissioner Baird, who related the events in his annual report, was not particularly concerned: "Many of the species … were well adapted to the waters of the river, but of course not the tautogs, lobsters or oysters."[17]

By accident or design, the effect of moving eggs, juveniles, and adult fish from place to place was an extensive experiment in rearranging the biological contours of riverine ecosystems. Fish adapted for life in particular rivers were moved to other rivers, without full comprehension of the effects. The establishment of hatcheries in the late-nineteenth century outpaced the development of scientific understanding. Not until the late 1930s did biologists agree that salmon populations had distinct migratory runs and that they returned to their natal streams to spawn—long after many millions of fish had been stocked in rivers foreign to them. As with unreliable knowledge of ecosystems, so with uncertain classifications of fish. Contrary to the Commission's report, for example, species of Pacific salmon (*Oncorhynchus*) belong to a different genus than Atlantic salmon (*Salmo*). Based in part on a nascent science of fish biology and in part on pragmatic efforts to distribute fish, a new mosaic of species began to enter river systems, as U.S. hatcheries sent eggs, juveniles, and adult fish around North America and beyond, to Australia, New Zealand, Chile, Argentina, France, and Germany. In return, U.S. hatcheries often received eggs and fish from distant sources. Improving existing stocks, introducing new stocks, hatching and rearing fish, and distributing the results became central features of fisheries management.[18]

The promise of hatcheries, and the belief that scientific and technological solutions could be devised to restore salmon and other fish stocks, received public backing and strong Congressional support. Many of the recognized differences between fish species and the environments that sustained them were thought not to matter. Fish could live in rivers almost anywhere, the argument went, given adequate food sources, compatible temperatures, and suitable spawning grounds. By the turn of the century, fisheries managers promoted not the control of human behavior but the development of fish culture as the preferred means to increase populations of fish. Producing endless supplies of domesticated fish was more palatable than enacting laws regulating humans in order to restore natural runs. The scene was set for the twentieth century, in which hatcheries and the bureaucracies that supported them became self-perpetuating, while salmon continued to decline.[19]

Twentieth-Century Salmon Restoration

Despite Atkins' considerable efforts, salmon extinction was still a concern at mid-century. In 1949, the Atlantic Sea-Run Salmon Commission, established to develop policies to improve salmon runs in Maine, made observations in its first report that echoed those of the prior generation of researchers. "Except for the small run remaining in the Penobscot, natural runs of salmon had become almost extinct by 1920. Only continuous stocking efforts kept small runs alive in the badly obstructed rivers of eastern Maine." By 1949, when the Atlantic Sea-Run Salmon Commission began publishing its reports, there was still reason for hope. The extinction in Maine rivers was not absolute. Wild salmon continued to spawn in very small numbers. River conditions had improved, and fisheries biologists were beginning to better understand the limits and possibilities of stocking programs. As the Salmon Commission's report acknowledged, not all of Maine's rivers "are permanently lost." In concert with stocking programs, other human interventions might forestall such loss. "Pollution abatement, fishway construction and the screening of water diversions" might prove sufficient to restore salmon to many rivers in Maine. With the cooperation of the U.S. Fish and Wildlife Service, and the Maine Inland Fisheries and Game Department, a new plan was developed to restore salmon runs.[20]

Recreational fishing was one driving force. A 1968 national review of hatcheries and sport fishing noted that

> Maine fishery habitat is capable of sustaining considerable angler pressure. An exception is Atlantic salmon. Populations of this species have been reduced through destruction of spawning and nursery areas by pollution, or restriction from these areas by dams, until survival of the species is questionable. A concerted effort is being made to restore the species by eliminating pollution and providing access over dams.

The anticipated result would be renewed sport fishing for salmon in the near future.[21]

Yet by the 1980s, environmental problems and inconsistent management practices persisted. The Atlantic Sea-Run Salmon Commission's 1982 report on the Machias River noted the contradiction between land use management and fisheries management:

> Timber and pulpwood stands in the Machias drainage have been clear-cut in recent years, and current harvesting practices can adversely affect the waters of the drainage. Roads and skidder trails built to reach harvestable stands can cause erosion problems. Even moderate persistent erosion can cause siltation of desirable stream bottom areas necessary to the production of juvenile salmonids. Deforestation of large tracts of land within a watershed can also result in a destabilization of water flows, with increased runoff during the spring and following heavy rains. The water retention capabilities of the forest soils may be reduced, and result in lower flows during critical dry months.[22]

The Commission's report on the Narraguagus River made similar observations. Following a description of the effects of fires in 1840, 1880, 1884, 1903, and 1916, which "removed much of the organic soil and ground cover" in the drainage, the authors of the report noted,

> The changes in the fisheries resources of the region were less visible but were just as profound. Sawmill and driving dams altered water flows and deposited refuse in streams and lakes. The stocks of salmon, shad and alewives were

depleted as spawning and nursery areas were inundated and migratory routes barred by impassable structures.[23]

But the historical effects of fires, dams, and mills proved to be only part of the problem. In the 1980s, Narraguagus watershed lands were managed for blueberry production and for forest products. The effects on salmon were understudied but potentially extreme. "The potential for serious impacts upon the fish resources is very real since insecticides and herbicides are used extensively by these industries," the Commission's report noted. Moreover, as with the Machias watershed, "Forest cutting practices can cause serious erosion and/or siltation problems in the river."[24]

Despite the multitude of environmental problems and the difficulties coordinating land and river management strategies, salmon were beginning to return to Maine rivers in higher numbers. River-specific management plans, stocking efforts, and other improvements to Maine rivers appeared to be successful. In addition, for reasons that are still unclear, salmon apparently survived at higher rates in the ocean than in prior years, thus returning in greater numbers to their home streams to spawn. By the mid-1980s salmon populations had rebounded in Maine, with returning fish numbering in the thousands. Recreational fishers were delighted and took somewhere between 1,000 and 2,000 Atlantic salmon annually from Maine rivers.[25]

By some measures, the results of more than a century of hatchery efforts were impressive. In the last quarter of the nineteenth century and throughout the twentieth century, fisheries managers experimented with different hatchery techniques and stocked salmon and other fish during different stages of their life cycles. In Maine rivers alone, between 1870 and 1995, they hatched and released 67,866,560 salmon fry, 14,057,325 parr, 14,088,430 smolts, and 2,624 adult salmon.

The fry stage refers to young free-swimming salmon, after they have absorbed their egg sacks. The parr stage, named after the bar-like markings that appear on juvenile salmon, refers to the fresh water life of salmon, before they migrate to the ocean. The smolt stage refers to the point when salmon undergo a variety of physiological changes in preparation for their migration to the ocean where they spend several years feeding, before returning to spawn. All told, over more than a century and a quarter of effort, 96,012,315 young salmon were stocked in Maine rivers—nearly 100 million fish.[26]

By other measures, however, the results were less than impressive. The hard work and good intentions of many people to restore salmon runs may have been too little, too late. Or they may have been simply ill-advised, perhaps targeting the wrong causes of declines and celebrating what turned out to be short-term restoration effects.

By the end of the 1980s, salmon populations again markedly declined. Marine survival rates dropped by as much as 80 percent. Returns were suddenly so low that in 1991 the U.S. Fish and Wildlife Service listed five Maine rivers as Category 2 under the Endangered Species Act. This category has since been removed from the Act, but at the time meant that the Service was sufficiently concerned with the long-term survival of Atlantic salmon to begin gathering biological information about the species, in anticipation of listing it as threatened or endangered.[27]

In the early 1990s, after more than 100 years of stocking salmon from other rivers, the federal strategy shifted to stocking river-specific fish. It became increasingly apparent that to foster self-perpetuating salmon required hatching and releasing fish in the same river that their parents were from. Such fish returned in greater numbers than did fish whose parents came from other rivers.

For Atlantic salmon, the nineteenth-century experiment in human-facilitated fish migration had failed. Salmon from distant rivers have never returned to spawn in sufficient numbers to obviate the need for further hatchery work.[28]

To make and distribute fish required the development of a professional group of fish managers, and state and federal bureaucracies to fund and organize their efforts. The simple fact that we know the absolute numbers of salmon stocked in Maine rivers over a hundred-year period indicates a bureaucracy at work. By the end of the twentieth century, the second largest budget line for the U.S. Fish and Wildlife Service was for hatcheries, the largest was for maintaining wildlife refuges. In New England alone, $200 million was spent on hatchery operations over a recent 20-year period. With origins in the late-nineteenth century, hatcheries and the bureaucracies to support them had become institutionally secure by the second half of the twentieth century. Yet salmon restoration proved elusive.[29]

Continuing Salmon Threats

The pattern of Atlantic salmon decline is worldwide (Figure 3.1.). Wild Atlantic salmon have disappeared from more than 300 river systems in North America and Europe, and many more river systems will probably lose their salmon populations in the next few decades. How to improve the situation is far from clear. Localized stocking efforts are only part of the solution and may be part of the problem.

FIGURE 3.1 Map of the Ocean Distribution of Atlantic Salmon

Source: wiki commons. https://commons.wikimedia.org/wiki/File:Distribution_of_Atlantic_salmon.svg.

There is some evidence that the stocked fish diminish the survival chances of wild fish. In addition, salmon live complex lives during which they pass through many ecosystems and many political jurisdictions. Along the way, humans harvest salmon and make extensive use of salmon habitat and have done so in North America for more than two centuries.[30]

Changes to the industrial landscape of Maine continued to affect salmon. In the 1980s, salmon farming became a booming industry, with escaped salmon possibly interbreeding with wild salmon, with unknown but potentially disastrous consequences for wild stocks. For a time, more than ten million farmed salmon lived out their lives in pens off the Maine coast. Like other domesticated animals, farmed salmon have been selected for desirable characteristics such as fast weight gain, high survival rates, and high market value. Genetic manipulation, to further improve stocks, is a key component of the industry. In Maine, preferred stocks of farm-raised fish were a blend of strains from the Penobscot River (Maine), the Saint John River (New Brunswick), and Scotland. The Scotland fish, called Landcatch, is composed of various Norwegian strains of salmon from 41 different rivers. As genetic blends, farmed salmon in Maine contain 30 to 50 percent European genes.[31]

Containment became a major issue, since even a small percentage of farmed escapees could dominate wild salmon habitat, contribute genetic characteristics unsuited for life in the wild, or transmit disease. One viral disease, infectious salmon anemia, threatened farmed salmon.

The virus was first detected in Norwegian and later in Canadian fish. By 1999, it had not been detected in U.S. waters. Two years later, however, one million Maine fish had become infected. By January of 2002, the problem became acute and the State of Maine ordered all farmed salmon in Cobscook Bay destroyed, approximately 1.5 million fish. The apparent solution to the viral disease was to destroy all infected or potentially infected fish and to let the pen sites lie fallow for a season or more. The virus, denied its host, would then be flushed out by normal tides and dissipate. Pens, boats, and associated equipment were removed from the water and sterilized. In 2002 the U.S. Department of Agriculture released $16.6 million to pay for the cleanup.[32]

Beyond Maine's jurisdiction, an open ocean fishery decimated stocks of wild salmon. Atlantic salmon from North American rivers continued to be harvested, specifically in the West Greenland fishery. In 1971, the West Greenland salmon catch was 2,689 metric tons. The next year, a quota system of management was introduced, which set allowable catch limits at 1,100 metric tons. Nine years later, the quota was roughly equivalent, at 1,265 metric tons. By the end of the 1980s, however, fishing fleets could not meet their allowable quotas because salmon had become too scarce. In recognition of depleted stocks of Atlantic salmon, the quota was reduced to 840 metric tons.

The collapse of the West Greenland fishery was indicative of a larger pattern of exploitation. Worldwide, the total catch of Atlantic salmon dropped from 4 million fish in 1977, to 2.6 million fish in 1987, to 800,000 fish in 1997—an 80 percent drop over 20 years.[33]

Fearing the complete extirpation of Atlantic salmon in North America, the National Fish and Wildlife Foundation, under the directorship of Amos Eno, raised $5 million to buy out the West Greenland fishery in 1993 and 1994, which effectively closed the fishery and resulted in a modest rebound of salmon. The international North Atlantic Salmon Fund, directed by Orri Vigfusson, participated in the buyout, and continued to purchase fishing quotas in Greenland and other threatened fisheries throughout the world. The North Atlantic Salmon Fund estimates that 70,000-90,000 fish were saved each year by the buyout.

In October of 1994, however, Vice President Al Gore apparently told Vigfusson that the U.S. Fish and Wildlife Service, and the National Marine Fisheries Service, did not support continued buyouts of the West Greenland salmon fishery. Gore argued that the North Atlantic Salmon Conservation Organization's (NASCO's) quota reductions averted any need for buyouts, that buyouts in any case benefited Canada and Europe, not the United States, and that halting the high seas intercept fishery would undermine U.S. Fish and Wildlife programs in New England while potentially compromising efforts at salmon restoration. The United States would not support the cessation of the high seas Atlantic salmon fishery.[34]

Negotiations to buy out the West Greenland quotas stalled in 1995, and the fishery reopened with a quota set at 174 metric tons, but only 92 metric tons were caught. From 1998 to 2000, Greenland restricted the catch to supply internal consumption only, to a maximum of 20 metric tons. NASCO subsequently negotiated an agreement between the United States, Canada, and Denmark (on behalf of Greenland) to limit the commercial take of Atlantic salmon in 2001 to 28 metric tons, or about 11,000 fish, with the fishery open only for seven days in mid-August. NASCO hoped that the limit would forestall the continuing decline of salmon in North American rivers. The allowable take, however, was an increase from prior years. Scientists, policymakers, and commercial fishers had not come to a consensus about what such limits should be, despite decades of declining salmon stocks.

Official U.S. policy since 1969 calls for the complete ban of all commercial take of Atlantic salmon, a ban supported by many environmental organizations. Yet, the United States has done precious little to halt the high seas intercept fishery. In 1988, the National Marine Fisheries Service disallowed commercial take of Atlantic salmon in federal waters, from 3 to 200 miles offshore. However, the Clinton administration did nothing to stop the high seas salmon fishery, where wild salmon continue to be harvested outside of U.S. federal waters, and the George W. Bush administration followed suit. Later administrations appeared mildly interested at best.[35]

The ups and downs of the fishery, characterized by inconsistent management practices, indicate the difficulty governments have with developing rational fisheries policies. As the 1996 North Atlantic Salmon Fund Progress Report noted, "Usually governments lack the political powers to implement sensible and sustainable management of fisheries." Groups such as the National Fish and Wildlife Foundation, North Atlantic Salmon fund, and the Atlantic Salmon Federation, with sufficient economic clout to effect change outside of the usual governmental channels, may hold the key to improving the world's fisheries. In contrast to the hundreds of millions of federal dollars spent on hatcheries, a fisheries buyout is a relatively inexpensive management strategy.[36]

"Maine Salmon Is Extinct"

The social power of science as a persuasive force was used on all sides of the debate to bolster claims about the status of salmon and to promote particular views of appropriate human/salmon relationships. Under the Endangered Species Act, one key question was to determine whether a "Distinct Population Segment" (DPS) of Atlantic salmon exists in Maine rivers. Did salmon from a group of seven rivers and one tributary in Maine—the Dennys, East Machias, Machias, Pleasant, Narraguagus, Ducktrap, and Sheepscot Rivers, and Cove Brook—form a population of aboriginal fish, with characteristics that differentiate them from other aboriginal salmon populations in Canada and Europe? The Services said yes and referred to the few remaining wild salmon in these rivers as comprising a Gulf of Maine DPS. Other Maine rivers, such as the Penobscot, may have contributed to this distinct population of salmon, but the genetic and morphological bases for making that claim had not been conclusively demonstrated. More studies were needed.[37]

According to such critics as the State of Maine and the Maine Salmon Rescue Coalition (a group of Maine businesses), the Services had not demonstrated a distinct population for the rivers in question. No such population existed. The Services' science was conceptually flawed, logically inconsistent, and factually in error. A DPS required some level of isolation. The rivers flowing into the Gulf of Maine had not been isolated but had been subject to massive stocking efforts that included fish originating from rivers outside the Gulf of Maine. In addition, the percentage of wild fish that do not return to their natal streams to spawn, but instead swim up some other nearby river, was much higher than the Services argue—not two percent, but closer to 16 percent—which calls into question the idea of isolation. Perceived genetic differences may also be the result of genetic drift often found in small populations, which results in the fixing of deleterious genes and the general lowering of the fitness of individual organisms, measured in terms of relative reproductive potential.[38]

Since the Endangered Species Act requires the best science of the day, the scene was set for protracted arguments about what the best science was, who practiced it, and whether the results were conclusive or equivocal. When scientists disagree, as they often do in discussions of the structure and functioning of complex ecosystems, policy decisions concerning those ecosystems become especially contentious. Courts are frequently approached to resolve any resulting policy disputes. In the context of law, fishery biologists, geneticists, hydrologists, and other scientists become expert witnesses and their work enters a new social arena—the venerable, cantankerous, highly ritualized, and often contradictory theater of proof and counterproof found in most courtrooms. In this arena, scientific work is used in novel ways, subject to, even subordinate to, the discourse of law, in which it is not science but its ability to persuade nonscientists that is at issue.

The State of Maine preferred to keep salmon out of the courts. Litigation, former Governor Angus King argued, benefited lawyers, not salmon. In addition, Maine did not want its industries—agriculture, salmon farming, timber harvesting—to be burdened by a listing of Atlantic salmon under the Endangered Species Act. The fear was that, simply by filing lawsuits, private citizens or environmental groups could use the Act to force these and other industries to make prohibitively expensive changes to their current practices. Maine preferred a cooperative arrangement between industry, state bureaucracies, federal bureaucracies, and nongovernmental environmental groups. Such cooperative work in fact had begun in 1995. Despite considerable odds against timber companies, salmon farmers, blueberry growers, environmental advocacy groups, and government officials managing to find common ground, they collectively made an impressive effort to develop ways to improve salmon habitat, which eventually resulted in a federal decision not to list salmon as threatened or endangered.

Former Secretary of the Interior Bruce Babbitt was effusive in his praise of the collaborative effort. In a 1997 press release, he suggested that the cooperation between all parties was "a new chapter in conservation history." Babbitt saw both environmental improvements and economic benefits flowing from the novel relationships between state and federal agencies and public and private economic interests: "The governor [of Maine] showed great leadership in forging this collaboration, which will continue to enhance the ecology and economy of the state for years to come." More than this, the collaborative effort to improve Maine rivers will "stand as a model for the nation."

On December 15, 1997, at a public ceremony to celebrate the collaborative restoration effort, Babbitt noted:

> The announcement today is short and sweet by joint agreement of the National Marine Fisheries Service, in the Department of Commerce, and the Fish and Wildlife Service, in the Department of Interior. We are here gratefully and happily to say to the people of the great State of Maine the petition that lists the Atlantic Salmon is hereby withdrawn. Yeah, I kinda thought that would be a crowd pleaser! But it didn't just happen. This happy event today is the combination of a lot of work by some very determined people led by your good Governor who several years ago sat down with Molly Beattie, the Director of the Fish and Wildlife Service, and the people from the National Marine Fisheries and made a simple point, and that was that the protection of the Atlantic salmon ought to be worked out on the ground under the Governor's leadership by the affected people in the state agencies and the conservation organizations of the State of Maine. The Governor's pitch to us then and now was very simple. He says, "Rather than setting up the inevitably antagonistic form of federal regulation we can, working together, buy all of the stake holders into a plan to protect this fish because the people of Maine have a deep and abiding love for their land and their resources and this salmon," and that of course is what leads us to the work product today,

which is the Conservation Agreement which has been put together under the Governor's direction, which is the substitute for the regulatory action of listing and which substitutes precisely because by its terms it removes the threat that could've caused the listing. And I would say in conclusion that this is a big win for the people of Maine.[39]

A sigh of relief came from those in Maine who did not want to be constrained by an endangered species listing. The heavy work of improving local salmon habitats would continue under state leadership and with strong local participation.

For their parts, the U.S. Fish and Wildlife Service and the National Marine Fisheries Service recognized an obligation to list Atlantic salmon as endangered, despite any potential economic difficulties or benefits that result from the listing, provided there is appropriate scientific justification.

The Act, after all, is intended to preserve species on the edge of obliteration. It is not intended to allow industries, states, or the federal government to nudge species over the edge.

The Act specifically disallows economic considerations from entering into the decision to list. But it does allow a consideration of state and local efforts to preserve a species as a determining factor. If state and local efforts are sufficient, then a federal listing is unwarranted. The improvements to salmon habitat, laid out in a five-year state plan, could indeed stand as a model to the nation, as Babbitt foresaw, while the nation, in the guise of the Endangered Species Act, could be kept out of Maine.

The state plan was impressive. It endeavored, with the cooperation of all affected parties, to (1) improve fish management techniques; (2) restore degraded habitats and ensure future habitat integrity; (3) provide more comprehensive salmon protection; (4) develop new outreach organizations for public education; and (5) effectively enforce existing regulations.

By 1999, Maine had spent $1 million to implement its plan and had earmarked another $1 million for the effort. Fish weirs had been built to trap and count returning wild fish and escaped farmed fish. A long-term river-specific fry stocking plan was developed. Habitat assessments of critical spawning and nursery areas were completed. Water use management plans were being drawn up. One hundred and five beaver dams were removed, and other improvements to fish passages were made. New, albeit voluntary, codes to limit salmon from escaping from farms were devised. Recreational fishing for salmon was banned entirely. Watershed councils were formed for each river in question. New code enforcement officer training programs were developed, which highlighted the Maine Atlantic Salmon Conservation Plan. The State of Maine appeared to be fully behind the effort to restore salmon habitat.[40]

It was not to be. The environmental advocacy groups Trout Unlimited and the Atlantic Salmon Federation believed that the Maine plan was inadequate to preserve the species, mostly because Maine had failed to allocate sufficient funds for the task, even though on paper the Maine plan was worthy of admiration. As a practical matter, these environmental organizations argued, the plan was doomed to failure, with salmon following close behind. Trout Unlimited and the Atlantic Salmon Federation filed a lawsuit suit for an emergency endangered listing for Atlantic salmon.

The Services, without acknowledging that they were contemplating reversing their earlier ruling, disagreed that an emergency listing was warranted and presented to the court their rationale in two lengthy declarations. The process of determining the status of Atlantic salmon was well underway, they argued, and did not need emergency listing.[41]

The Services subsequently published a lengthy status report on Atlantic salmon in July 1999, followed in November by a proposed rule that reversed their earlier finding and determined instead that Maine Atlantic salmon was an endangered species in need of federal protection. In the July report, the Services concurred with Trout Unlimited and the Atlantic Salmon Federation about the adequacy of the State restoration plan, but for different reasons. The Services believed that remnant populations of Atlantic salmon were of special concern. Genetically pure populations of river-specific salmon probably no longer existed, given the stocking practices over the last century and interbreeding with escaped salmon from coastal farms. Nevertheless, the continued presence of river-specific salmon in their native habitat indicates, as the report on the status of Atlantic salmon argues, "that important heritable local adaptations likely still exist." In other words, the few native salmon that continued to spawn in Maine rivers were adapted to life in those rivers, and nowhere else. They were unique to those rivers. Under the Endangered Species Act, efforts should be made to ensure their continuing presence and to avoid their complete extirpation in U.S. rivers.[42]

But this was precisely the problem. Some scientists argued that the extirpation has already happened, and consequently native salmon no longer existed in Maine rivers—a position that the state of Maine and many Maine businesses adopted, for their own purposes. If there were no native salmon then the Endangered Species Act cannot apply: there was nothing to apply it to.

Addressing in June of 2001 a National Academy of Sciences (NAS) committee convened to study the issue, Governor King of Maine baldly stated, "Maine salmon is extinct." In an earlier talk, Governor King was similarly concise in his assessment of the appropriateness of the Endangered Species Act. "The fish in question," he said, "are neither endangered nor a species."[43]

One central policy problem, which King seems to have solved, was which set of scientists to believe, those who argued Maine salmon as distinct stocks of river-specific fish were extinct, or those who argued that they were nearly extinct. If extinct, then Maine was relatively free to adopt its own rules, unencumbered by the federal Endangered Species Act. If not, then the act clearly applies, and Maine, and the businesses operating in Maine, must take appropriate steps to bring all human/salmon interactions into compliance.

In his June address to the NAS committee, King sounded as if he had been transported from the Progressive Era. He asserted that, with adequate science, public policy decisions were self-evident. Given the complexity of the ecosystems salmon inhabit, however, many scientific issues remain unresolved or only partially resolved.

In addition, biologists recognize different kinds of extinction. These differences are rarely articulated in public policy debates. Local extinction refers to the displacement of a species from a relatively small area. Regional extinction refers to the displacement of a species from some large part of its historical range. Global extinction refers to the complete disappearance of a species. These are spatial definitions. There are also functional definitions. Functional extinction refers to circumstances in which "a species is so reduced in abundance that it no longer plays a quantitatively important role in the energy flow or the structuring" of an ecosystem. Commercial extinction refers to the reduction of a species by overhunting or overfishing to the extent that exploitation ceases or is significantly curtailed.[44]

As a species, Atlantic salmon was clearly not extinct in Maine. At the time of the endangered listing debates, more than ten million lived in pens off Maine's coast. But these were not salmon native to Maine's rivers.

Wild Atlantic salmon have been subject to local and regional extinctions in New England and have nearly become commercially extinct. They are also probably functionally extinct, but the role of wild salmon in the structure and functioning of the riverine, estuary, and ocean environments in which they live is not clear, nor for that matter is the functional role of stocked or farmed salmon clear.[45]

Despite Governor King's optimistic assessment of the value of science, policies concerning salmon were difficult to craft if the criteria were solely scientific. Vociferous debates often resulted. Reacting to the Services policy shift, Senators Olympia Snowe and Susan Collins both came to the defense of their home state.

Collins implicated the political nature of the policy process itself as central to the decision to list salmon as endangered. "Nowhere does the ESA permit a listing decision to be driven by a shift in Service policy or a national interest group's lawsuit meant to force a listing occur," she wrote. "Yet, it appears these sorts of motivations may underlie the Services' decision to abandon the [state] Plan." Collins argued that inadequate science, flawed logic, and bad policy combined to produce the listing.

Senator Snowe, in her statements at the Salmon ESA Listing Proposal Hearing, on January 29, 2000, similarly expressed great skepticism about the basis for the listing.

> When I look at the totality of this process—the inability to explain how a listing can achieve what a fully implemented State Plan can't … the reliance on inconsistent and unproven science … the reluctance to institute existing fisheries management tools … the only conclusion I can come to is that your [the Services'] decision was driven solely by the fear of losing the lawsuit brought against you.

Snowe further argued that the effects on Maine could be enormous, especially if the $68 million-a-year salmon farm industry, the $100 million-a-year blueberry industry, and others were all open to lawsuits if any of their practices were perceived, in the language of the Endangered Species Act, to "harass, harm, pursue, hunt, shoot, wound, kill, trap, capture, or collect" wild salmon, "or to attempt to engage in such conduct."[46]

The Final Rule

The final rule listing Atlantic salmon as endangered took effect on December 18, 2000. The Services determined, based on what they considered to be the best available scientific evidence, that a distinct Gulf of Maine population of wild Atlantic salmon existed and was in danger of extinction. This population included both wild and river-specific hatchery reared fish that had wild parents from Maine rivers. The Fish and Wildlife Service built an elaborate $12 million hatchery facility at Craig Brook, near the site of Charles Atkins' nineteenth century hatchery operations, to produce river-specific fish and to thereby assist in the recovery of salmon populations. If these river-specific hatchery fish spawn in the wild, they will count toward removing the Gulf of Maine population of Atlantic salmon from the endangered species list. Although no genetically pure forms of native fish exist, present populations were nevertheless descendants of aboriginal fish. These populations are essential to conserve because they "represent the remaining genetic legacy of ancestral populations that were locally adapted to the rivers and streams of the region." Preserving wild fish and augmenting them with river-specific hatchery reared fish ensured, or attempt to ensure, that this important genetic legacy does not disappear.[47]

In their decision to list Atlantic salmon as endangered, the Services cite what appeared to be failures of Maine policy to implement appropriate and effective restoration strategies. They note that between 1997 and 1999, a series of problems remained and new ones emerged. These include new diseases, aquaculture escapees, little or no progress in resolving problems with aquaculture practices, low survival of juvenile salmon within particular rivers, declining adult returns, recreational fishing problems, and water withdrawals from salmon bearing rivers.

With the possible exception of low juvenile salmon survival rates, these are all problems of human/salmon relationships. They are also problems that are part of a long history of failed human/salmon interactions, a history that caused the demise of wild Atlantic salmon in rivers throughout New England. Despite the listing and its consequences, these problems may prove fatal for populations of wild salmon in Maine. Governor King's assertion, "Maine salmon is extinct," may yet prove to be true.

The final rule identified 11 potential human activities that may violate the Endangered Species Act. These include (1) recreational or commercial fishing, accidental bycatch, and illegal fishing; (2) the escape of non-North American salmon from hatcheries within the range of the endangered salmon; (3) the escape of farmed salmon from marine cages or freshwater hatcheries; (4) failures to adequately guard against introducing diseases to the endangered population; (5) operating salmon farms in a way that degrades water quality or ocean-floor habitats; (6) polluting water that supports endangered salmon with "toxic chemicals, silt, fertilizers, pesticides, heavy metals, oil, organic wastes or other pollutants"; (7) obstructing migration routes; (8) destroying or altering habitat, for example, by "instream dredging, rock removal, channelization, riparian and in-river damage due to livestock, discharge of fill material, operation of heavy equipment within the stream channel, manipulation of river flow"; (9) violating permits allowing withdrawal or discharges in salmon habitat; (10) pesticide or herbicide use, regardless of whether such use violates or follows label restrictions; and (11) any handling or collecting of salmon that is unauthorized.

The Services foresee these activities as potential violations of the Endangered Species Act but note that other unforeseen activities could similarly violate the Act. With exceptions granted under the Act, prohibitions against harming, killing, selling, possessing, transporting, or delivering Atlantic salmon, "apply to all individuals, organizations and agencies subject to U.S. jurisdiction."[48]

With this ruling, the control of nature became quite explicitly the control of human beings. The longstanding emphasis on hatcheries as the preferred means to rectify human-caused damage to salmon populations and their habitats has been supplemented with, perhaps surpassed by, an emphasis on regulating human behavior, including all commercial, state, and federal activities. With exceptions granted by federal permit, all human actions affecting salmon and salmon habitat under U.S. jurisdiction must now comply with the Endangered Species Act.

Recovery at this point becomes a major issue. Federal endangered species recovery programs attempt to restore nature, but centuries-old human exploitation and manipulation of salmon and salmon habitat may render such attempts chimerical at best. In addition, recovery programs have been historically under-funded, a trend the George W. Bush and later administrations perpetuated.

The Fish and Wildlife Service defines recovery as

> the process by which the decline of an endangered or threatened species is arrested or reversed, and threats to its survival are neutralized, so that its long-term survival in nature can be ensured. The goal of this process is the maintenance of secure, self-sustaining wild populations of species with the minimum necessary investment of resources.

In 1990, 581 species were listed as threatened or endangered. Ten years later, 1,205 species were listed. By 2022, approximately 1,669 were listed as threatened or endangered in the U.S. The goal of all of these listings is to "restore listed species to a point where they are viable self-sustaining components of their ecosystems, so as to allow delisting."[49]

Environmental Values and the Policy Process

The Services are required to solicit and seriously consider local concerns in the development of endangered species policies and regulations. Public comment on proposed endangered status is a formal mechanism that informs the Services of local interests, concerns, and values. The comment period for the proposed endangered status of Gulf of Maine salmon was open from November 17 to February 15, 1999, which the Services twice extended. Over this time, the Services received more than 200 written comments. In addition, three public meetings were held in Maine in late-January and early-February 2000. In aggregate, more than 1,000 people attended.

In their final ruling, the Services responded to public comments, most of which focused on the questionable accuracy of the Services' scientific work. Since endangered status is presumably founded on scientific assessment, it is understandable that many public comments addressed issues of salmon genetics, life history, morphology, the effects of stocking, and the assumptions underlying the idea of a distinct population segment.

The principal question was whether salmon in a group of Maine rivers form a remnant population of aboriginal fish. The primary evidence comes from genetic studies and the continuity of wild fish populations in those rivers despite massive stocking over 130 years. The issue thus becomes narrowly defined as one resolvable by genetics. The ensuing debates made it appear as if the main values involved in salmon restoration were scientific values of objectivity, accuracy, and replicability. Analyses of microsatellite and mitochondrial DNA diversity of *Salmo salar*, which purport to demonstrate the genetic distance between river-specific fish, emerged as key elements in the debate. Those with the best science on their side won.

According to the National Academy of Sciences (NAS), the Services had the most compelling evidence. NAS organized a committee of fisheries experts comprised of 13 scientists from Norway, Canada, and the United States. The committee reviewed the available literature, conducted two fact-finding trips in Maine, and heard conflicting arguments from biologists, geneticists, environmental activists, industry representatives, and state officials about the history and status of salmon. In January 2002, the committee released its interim report, "Genetic Status of Atlantic Salmon in Maine," which essentially agreed with the federal listing. The report concluded that remnant populations of aboriginal salmon persist in Maine rivers in small numbers.[50]

But the scientific debate, necessitated by the requirements of the Endangered Species Act, obscured other debates and hid other values. Questions of economic harm or benefit were summarily dismissed. The Act precludes consideration of them. Questions of proper governance were likewise dismissed. If the State of Maine failed to forestall the imminent extinction of Atlantic salmon, then the Services clearly had the authority to determine endangered status and to implement recovery programs.

Questions of international scope, concerned with the open ocean intercept fishery, were relegated to international organizations that did not always act efficiently or wisely. In any case, such organizations could be ignored if they proposed solutions contrary to the interests of the U.S.

Scientific work becomes polarized in such contexts precisely because, under the Act, consideration of non-scientific issues is disallowed. Participants in listing debates were therefore forced to use science to promote political ends. The "Comments of the Maine Salmon Rescue Coalition," a group of Maine businesses, for example, deemed the Services' scientific work to be deeply flawed. They said the Services' scientific assessment was based not on any biological rationale but on political expediency that was "driven by the Services' desire to stake out as expansive a range as possible for Services' primary jurisdiction over salmon restoration." According to the Coalition's critique, policy hides behind faulty biology.[51]

The State of Maine similarly debated biology when the central issue was not biological but political, state vs. federal oversight. Maine characterized the Services' science as "replete with fundamental flaws in scientific technique, errors of assessment and interpretation, unaccountable omissions of relevant scientific literature, misrepresentation of scientific studies, inconsistencies with prior determinations … and frequent assertions that are unsubstantiated." In its comments on the proposed endangered status of salmon, the State of Maine attempted a point-by-point refutation of the Services' science. The State of Maine interpreted the available scientific evidence as indicating that the Services failed to demonstrate the existence of a distinct population of aboriginal Maine Atlantic salmon.[52]

Governor King's letter to the Services, which accompanies the State of Maine's comments on the proposed listing, emphasized the importance of state restoration efforts. He noted that such efforts effectively "harnesses the self-interest of a broad spectrum of constituents including landowners, volunteers serving on watershed councils, local municipal governments, interested private citizens, and representatives of the numerous industries that operate in the affected areas." Nullifying state efforts before they can be properly assessed may result in the loss of local cooperation. The Maine Salmon Rescue Coalition, comprised the industries King referred to, directly addressed this possibility. An endangered species listing "will destroy the voluntary private support essential for effective long-term restoration of Maine salmon runs." Moreover, the Coalition argued,

> The Services' capricious decision, at odds with all prior representations, will merely serve to teach private landowners and citizens the futility of working voluntarily with the Services. Landowners will learn that voluntary cooperation with the Services under a partnership merely serves to establish a regulatory floor, from which the Services will unilaterally construct additional regulatory mandates in the future.[53]

Values, in these examples, are not scientific values but social values of self-determination, cooperation, and local control. The larger environmental debate and the narrower genetic debate are in fact debates about proper governance. From the perspective of many businesses in Maine and articulated in the Coalition's critique, environmental values should preserve "traditional primary state wildlife management," cultivate "private landowner support needed for any effective restoration effort," and defend against "creeping federalization of wildlife management."[54]

Since the concerns associated with local control of local environments cannot explicitly enter the Services' biological rationale to list a species as endangered, the only recourse for those opposing the listing was to cast local concerns in terms of the scientific debate. In this way, local values enter policy in disguised form, masquerading as discussions of salmon genetics, morphology, and life history. While environmental policy turns on the outcome of genetic studies, the values that were of immediate concern to many people in Maine become lost in the scientific debate.

Wild Salmon

As a migratory species, Maine Atlantic salmon is difficult to manage. Salmon do not adhere to the political boundaries humans impose on the natural world. In decreasing numbers, they spawn in gravel beds in a few rivers in Maine, migrate to the sea, and travel through U.S. and Canadian waters to feeding grounds in the open ocean west of Greenland, to return after three or more years to start the process anew. Wild Maine Atlantic salmon maintain the natural cycle of anadromous fish, despite humans and their activities.

But in an important sense, Maine salmon are also the product of human boundaries and of the policies that have operated within them. Humans intervene at all stages of their complex lives. Hatcheries, salmon farms, intercept fisheries, dams, timber harvesting, blueberry production, recreational fishing—through these kinds of human activities, Atlantic salmon have become thoroughly intertwined with humans and with the products of human labor. The relationship has largely failed, but it has taken more than a century for the consequences to become clear and for policies to better target the human side of the relationship. In the process, science has become both the referee and the authority on debates about the fate of Atlantic salmon.[55]

By the end of September of 2002, four adult salmon had returned to the Dennys River, seven to the Narraguagus River, and none to the Pleasant River. "It's been a consistent downhill slide," noted one salmon biologist. Hope now rested on new river-specific hatchery protocols. Wild salmon can only be maintained with adequate hatchery science—a longstanding belief in fisheries management—and with limits to human behavior.

That same month, Craig Brook National Fish Hatchery was officially opened, with a public ceremony commemorating the occasion. Music, speeches, and various awards heralded the opening. In the event that the hatchery failed, as previous hatcheries have failed, it may be closed. Although optimistic about the future of salmon in Maine, Mamie Parker, northeast regional director of the U.S. Fish and

Wildlife Service, was clear about the results of the hatchery: "If at any given point, we feel we are not making a difference, I will pull the plug."[56]

Has it worked? By 2009, Atlantic salmon in the Penobscot, Kennebec, and Androscoggin Rivers were added for protection as part of the endangered Gulf of Maine Distinct Population Segment.[57] Numbers of Atlantic salmon, despite considerable human effort to restore the species, remained at historic lows. In 2018, 862 adult salmon returned in spawning runs to Maine rivers, most of them to the Penobscot River. The majority of these fish originated as hatchery-stocked smolts. Natural, non-human assisted salmon totaled 98 returned fish. "Abundance remains critically low," as a recent assessment report dryly notes.[58]

The goal is for 6,000 wild spawners to return to Gulf of Maine rivers, and to then phase out hatchery support for Atlantic salmon, at which point the plug will be pulled. In 2020, however, it remained unclear, also unlikely, that Atlantic salmon will become self-sustaining once again. If that occurs, limitations to human behaviors may be lifted.

Meanwhile, as wild salmon struggled and the endangered species listing played out, the ocean will effectively move inland, and salmon along with it. A new industrial facility—a vast salmon farm—is being built in Bucksport, Maine. This facility, owned by Whole Oceans, will produce farmed Atlantic salmon, perhaps as much as 20,000 metric tons annually, attempting to capture a share of the global Atlantic salmon market, which in 2019 stood at $10 billion.

Farmed salmon no longer need to be subject to a life from hatcheries to ocean pens, where they are fattened for human tables. With the ocean brought on land on the site of a defunct paper mill, whose industry helped bring about the demise of wild salmon, Atlantic salmon will be entirely landlocked. Or at least this version of engineered salmon will live out their lives without benefit of rivers, estuaries, or open oceans.[59] Perhaps, wild Atlantic salmon will eventually benefit, as their genetically engineered cousins are sequestered inland.

Salmon restoration efforts and failures have a long history in New England. Until recently, restoration has been associated with hatcheries, closures of recreational fisheries, and limits imposed on commercial fisheries, as well as with technological means to overcome human-caused problems such as overfishing, dams, pollution, and other effects of industry. With the endangered species listing of Atlantic salmon in place since 2000, the locus of restoration effort shifted to include any and all human behaviors that may adversely affect salmon populations and habitats under U.S. jurisdiction. The federal listing resulted from previous policy failures. Better focused on human behavior than earlier policies, the listing attempted to reverse the historical trend that leads to the regional extinction of wild Atlantic salmon.

Under the Endangered Species Act, the best science of the day can be used as a justification to regulate human behavior. Because of a century of policy failures, and the apparent inability of hatcheries to reverse those policy failures, salmon now provide the opportunity to regulate people. Biology thus becomes policy and the control of nature becomes the control of human beings.

Notes

1   The primary sources on the debate include, "Review of the Status of Anadromous Atlantic Salmon (*Salmo salar*) under the U.S. Endangered Species Act," 60 Federal Register 50530 (July 1999); "Proposed Endangered Status for a Distinct Population Segment of Anadromous Atlantic Salmon (*Salmo salar*) in the Gulf of Maine," 64 Federal Register 62627 (November 17, 1999); "Comments of the Maine Salmon Rescue Coalition in Opposition to the Proposal to List Atlantic Salmon as Endangered Under the Endangered Species Act" (April 12, 2000); and "Comments of the State of Maine in Opposition to Proposed Endangered Status for a Distinct Population Segment (DPS) of Atlantic Salmon in the Gulf of Maine, 64 Federal Register 62627 (November 17, 1999)," (April 14, 2000). The published and unpublished documents concerning the endangered species listing have been collected by the Services and are available for inspection at the National Marine Fisheries Service, Gloucester, Massachusetts, and at the U.S. Fish and Wildlife Service, Hadley, Massachusetts.

2   On definitions of relative wildness and for spawning figures, see Ed Baum, *Maine Atlantic Salmon: A National Treasure* (Hermon, ME: Atlantic Salmon Unlimited, 1997), 2, 12, 28. The number of eggs a female produce depends upon body weight, with an average of 800 eggs for each pound of body weight. On state vs. federal definitions, see "State Federal Disagreement Hinges on Defining 'Salmon'," *Kennebec Journal* (December 13, 2000).

3   For stocking figures, see Baum, *Maine Atlantic Salmon*. On the escape of farmed salmon, see Mary Anne Clancy, "100,000 Salmon Escape from Pens. Aquaculture Moratorium Sought," *Bangor Daily News* (February 23, 2001); Roberta Scruggs, "Salmon Escape to Controversy," *Portland Press Herald* (February 24, 2001).

4   The 1997 determination is at 62 *Federal Register* 66325 (December 18, 1997). The 1999 Status Review can be found at http://news.fws.gov/salmon/asalmon.html. The proposed rule is at 64 *Federal Register* 62627 (November 17, 1999). "Final Endangered Status for a Distinct Population Segment of Anadromous Atlantic Salmon *(Salmo salar)* in the Gulf of Maine," 65 *Federal Register* 69459 (November 17, 2000). The amended Endangered Species Act is found at 16 *U.S.C.* §1531 et seq. See Alton Chase, *In a Dark Wood* (Boston, MA: Houghton Mifflin, 1995), ch. 7, for a historical overview of the Endangered Species Act. Earlier federal documents on Atlantic salmon include 59 *Federal Register* 3067 (January 20, 1994); 60 *Federal Register* 14410 (March 1, 1995); 60 *Federal Register* 50530 (September 29, 1995). See also "Comments of the State of Maine."

5   Henry B. Bigelow and W.W. Welsh, "Fishes of the Gulf of Maine," *Bull. U.S. Bur. Fish*, 40:130–138 (1924); William C. Kendall, "The Fishes of New England: The Salmon Family, Part 2," *Memoirs of the Boston Society of Natural History: Monographs on the Natural History of New England* 9 (1935); George A. Rounsefell and Lyndon H. Bond, "Salmon Restoration in Maine," Atlantic Sea-Run Salmon Commission, Research Report No. 1 (Augusta, ME, 1949).

6   Charles G. Atkins, "On the Salmon of Eastern North America, and Its Artificial Culture," in *Report of the Commissioner for 1872 and 1873, Part 11, United States Commission of Fish and Fisheries* (Washington, DC: Government Printing Office, 1874), 289–290.

7   Atkins, "On the Salmon of Eastern North America, and Its Artificial Culture," 289–290. On Atkins' career, see Ed Baum, *Maine Atlantic Salmon: A National Treasure* (Hermon, ME: Atlantic Salmon Unlimited, 1997).

8   Anthony Netboy, *Atlantic Salmon: A Vanishing Species?* (Boston, MA: Houghton Mifflin, 1968), 326. For a deeper history, see Brian S. Robinson, George L. Jacobson, Martin G. Yates, Arthur E. Spiess, and Ellen R. Cowie, "Atlantic Salmon, Archaeology and Climate Change in New England," *Journal of Archaeological Science* 36 (October 2009).

9 Richard W. Judd, *Aroostook: A Century of Logging in Northern Maine* (Orono, ME: University of Maine Press, 1989), 48, 159. The quote is from lumberman John A. Morrison, testifying to the International Saint John River Commission. See also Richard G. Wood, *A History of Lumbering in Maine, 1820-1861* (Orono, ME: University of Maine Press, 1971), originally published in *Maine Studies* No. 33 (1935); David C. Smith, *A History of Lumbering in Maine, I 861-1960, Maine Studies No. 93* (Orono, ME: University of Maine Press, 1972); and Alfred Geer Hempstead, *The Penobscot Boom and the Development of the West Branch of the Penobscot River for Log Driving, 1925-1931* (self-published, 1975 repr. of 1931 ed.). For a discussion of earlier timber practices, see Charles F. Carroll, *The Timber Economy of Puritan New England* (Providence, RI: Brown University Press, 1973).

10 Netboy, *Atlantic Salmon*, 326; E.M. Stillwell, "Obstructions in the Rivers of Maine," *Report of the Commissioner for 1872 and 1873, Part II, United States Commission of Fish and Fisheries* (Washington, DC: Government Printing Office, 1874).

11 Joseph E. Taylor, III, *Making Salmon: An Environmental History of the Northwest Fisheries Crisis* (Seattle, WA: University of Washington Press, 1999), ch. 3.

12 Atkins, "On the Salmon of Eastern North America, and Its Artificial Culture," 250.

13 Atkins, "On the Salmon of Eastern North America, and Its Artificial Culture," 250–251.

14 Atkins, "On the Salmon of Eastern North America, and Its Artificial Culture," 263.

15 On the origins of salmon fish culture, see Taylor, *Making Salmon*, ch. 3. See also Spencer F. Baird, *Report of the Commissioner for 1872 and 1873. Part II*, for a description of the need for a state/federal division of labor. U.S. Commission of Fish and Fisheries (Washington, DC: Government Printing Office, 1874), lxxxiii. Baird, in his *Report of the Commissioner for 1873-4 and 1874-5, Part Ill* (Washington, DC: Government Printing Office, 1876), also provides background on the origins of U.S. fish culture. He notes (p. xviii) that Germany sent 250,000 salmon eggs to the United States in 1873.

16 Baird, *Report of the Commissioner for 1873-4 and 1874-5, Part Ill*, xv.

17 Baird, *Report of the Commissioner for 1873-4 and 1874-5, Part Ill*, xxxviii.

18 Taylor, *Making Salmon*, ch. 3.

19 Taylor, in *Making Salmon*, argues that "state and federal hatcheries became so fiscally dependent on fish culture that criticism became taboo" (p. 98). He suggests that a simple historical narrative, in which past mistakes were corrected and environmental understanding improved, obscures the more intricate historical interplay of science, technology, and politics that produced the emphasis on fish culture as the preferred management strategy.

20 Rounsefell and Boyd, "Salmon Restoration in Maine," 5, 21.

21 *National Survey of Needs for Hatchery Fish*, Bureau of Sport Fisheries and Wildlife Resource Publication 63 (Part II) (Washington, DC: Government Printing Office, 1968). See also Herb Hartman, ed, *The Atlantic Salmon in Maine, a Complex and Valuable Resource: A Special Report* (Ipswich, MA: Atlantic Salmon Federation, 1989); James E. Butler and Arthur Taylor, *Penobscot River Renaissance: Restoring America's Premier Atlantic Salmon Fishery* (Camden, ME: The Silver Quill Press, 1992).

22 James S. Fletcher, Richard M. Jordan, and Kenneth F. Beland, "The Machias River: An Atlantic Salmon River Management Report" (Bangor, ME: Atlantic Sea-Run Salmon Commission, 1982), 34–35.

23 Edward T. Baum and Richard M. Jordan, "The Narraguagus River: An Atlantic Salmon River Management Report" (Bangor, ME: Atlantic Sea-Run Salmon Commission, 1982), 19.

24 Baum and Jordan, "The Narraguagus River," 37.

25 Baum, *Maine Atlantic Salmon*, 119, and Appendix 2.

26 Baum, *Maine Atlantic Salmon*, Appendix 6.

27 On the listing of five Maine rivers as Category 2, see *56 Federal Register* 58804 (November 21, 1991).

28 In rivers in other parts of the world stocking programs also seem to be ineffective. See Arne Fjellheim and Bjorn-Ove Johnsen, "Experiences from Stocking Salmonid Fry and Fingerlings in Norway," *Nordic Journal of Freshwater Research* 75:20–36 (2001).

29 The development of bureaucracies to support fisheries management was of course part of a larger historical trend in the development of national bureaucracies. See Stephen Skowronek, *Building a New American State: The Expansion of National Administrative Capacities, 1877-1920* (Cambridge: Cambridge University Press, 1982). The gathering of statistics is one well-known function of bureaucracies. For early fish statistics, see "Statistics of the Fisheries of the New England States," U.S. Fish Commission Report (Washington, DC: Government Printing Office, 1900); "Statistics of the Fisheries of the New England States, 1902," Bureau of Fisheries Report (Washington, DC: Government Printing Office, 1904); and, "Statistics of the Fisheries of the New England States, 1905," Bureau of Fisheries Document No. 622 (Washington, DC: Government Printing Office, 1907). The $200 million figure, apparently for the years 1974-1994, excludes the construction costs of the new hatchery facility at Craig Brook; the figure is reported in the National Fish and Wildlife Foundation report, "Ending the Greenland Atlantic Salmon Fishery: The U.S. Position and Prospects for Funding" (November 15, 1994), reproduced as Appendix 22 in "Comments of the Maine Salmon Rescue Coalition."

30 On the decline of Atlantic salmon, see "The Status of Wild Atlantic Salmon: A River by River Assessment," World Wildlife Fund (May 2001). See also John Anderson, Fred Whoriskey, and Andy Goode, "Atlantic Salmon on the Brink," *Endangered Species Update* 17:15–21 (January/February 2000). On the possibility that stocked Pacific salmon may adversely affect survival of wild salmon, see Phillip S. Levine, Richard Zabel, and John Williams, "The Road to Extinction Is Paved with Good Intentions: Negative Associations of Fish Hatcheries with Threatened Salmon," *Proceedings of the Royal Society Bulletin* (June 7, 2001). During much of the twentieth century, paper mills dumped highly toxic chemicals such as dioxin into Maine rivers, poisoning the rivers and killing large numbers of fish. Agricultural and timber harvesting practices have resulted in pesticides and herbicides being flushed into rivers with each rainstorm. Blueberry growers remove water from salmon-bearing rivers to irrigate their berries. Dams have been placed at almost every drop on every river with sufficient flow to generate power. The effects on fish are still being studied. For an assessment, which has clear implications for the health of Atlantic salmon, see Kim Chamberland, Beth Ann Lindroth, and Blake Whitaker, "Genotoxicity in Androscoggin River Smallmouth Bass," *Northeastern Naturalist* 9:203–212 (2002).

31 See T. Gjedrem, H.M. Gj!ilen, and B. Gjerde, "Genetic Origin of Norwegian Farmed Atlantic Salmon," *Aquaculture* 98:41–50 (1991); Ed Baum, "History and Description of the Atlantic Salmon Aquaculture Industry of Maine," Division of Fisheries and Oceans. Canadian Stock Assessment Secretariat Research Document (DFO) 98/152 (1998). On genetic manipulation, see T. Reichhardt, "Will Souped Up Salmon Sink or Swim?" *Nature* 406:10–12 (2000). A ruling in U.S. District Court, District of Maine, prohibits any further stocking of salmon of non-North American stock or genetic strain in waters adjacent to Maine's coast. U.S. Public Interest Research Group, et al., v. Atlantic Salmon of Maine, LLC, Civil No. 00-151-B-C, and U.S. Public Interest Research Group, et al., v. Stolt Sea Farm, Inc., Civil No. 00-149-B-C. May 28, 2003.

32 On aquaculture see Rebecca Goldburg and Tracy Triplett, "Murky Waters: Environmental Effects of Aquaculture in the U.S." (Environmental Defense Fund, 1997); Rebecca

J. Goldburg, Matthew S. Elliott, and Rosamond L. Naylor, "Marine Aquaculture in the United States: Environmental Impacts and Policy Options" (Arlington, VA: PEW Oceans Commission, 2001); on Maine aquaculture see Philip W. Conkling, "Fish or Foul? Will Aquaculture Carve Out a Niche in the Gulf of Maine?" *Maine Policy Review* 9:12–19 (Fall 2000); Paul Molyneaux, "Down on the Farm," *National Fisherman* (September 2001); and "Aquaculture: Its Challenges for the Wild Atlantic Salmon, a Brief to the Senate Committee on Fisheries" (Atlantic Salmon Federation, February 29, 2000). On the interactions between farmed and wild salmon, see Department of Fisheries and Oceans, "Interaction between Wild and Farmed Atlantic Salmon in the Maritime Provinces," DFO Maritimes Regional Habitat Status Report 99/1E (1999). For a popular-press account of the farming of Atlantic salmon in British Columbia, see Bruce Barcott, "Aquaculture's Troubled Harvest," *Mother Jones* (November/December 2001); see also Brian Harvey and Misty MacDuffee, eds., "Ghost Runs: The Future of Wild Salmon on the North and Central Coasts of British Columbia" (Victoria: Raincoast Conservation Society, 2002). For problems of containment, see Fred Whoriskey, "Causes of the Escaped Farmed Atlantic Salmon from Sea Cages in British Columbia and North America," International Council for the Exploration of the Sea, North Atlantic Salmon Working Paper (March 2001), and Freya Keyser, Brendan F. Wringe, Nicholas W. Jeffery, J. Brian Dempson, Steven Duffy, and Ian R. Bradbury, "Predicting the Impacts of Escaped Farmed Atlantic Salmon on Wild Salmon Populations," *Canadian Journal of Fisheries and Aquatic Sciences* 75:4 (April 2018). See also Deborah A. Bouchard, K. Brockway, C. Giray, W. Keleher, and P.L. Merrill, "First Report of Infectious Salmon Anemia (ISA) in the United States," *Bulletin of the European Association of Fish Pathologists* 21:86–88 (2001); and "Declaration of Emergency Because of Infectious Salmon Anemia," 66 *Federal Register* 65679 (December 20, 2001). The virus apparently has a variety of strains. See Rachael J. Ritchie, M. Cook, K. Melville, N. Simard, R. Cusack, and S. Griffiths, "Identification of Infectious Salmon Anaemia Virus in Atlantic Salmon from Nova Scotia (Canada): Evidence for Functional Strain Differences," *Diseases of Aquatic Organisms* 44:171–178 (April 2001). For news-paper accounts, see Andrew C. Revkin, "Virus Is Killing Thousands of Salmon," *New York Times* (September 7, 2001); and John Richardson, "Virus Spells End for Farmed Salmon," *Portland Press Herald* (January 16, 2002).

33 For catch statistics see "Report of the Working Group on North Atlantic Salmon, Part Two," Advisory Committee on Fishery Management, International Council for the Exploration of the Sea (April 2001).

34 National Fish and Wildlife Foundation report, "Ending the Greenland Atlantic Salmon Fishery," 2.

35 On marine issues and the open ocean salmon fishery, see Fred G. Whoriskey and K.E. Whelan, eds., *Managing Wild Salmon: New Challenges-New Techniques,* Proceedings of the Fifth International Atlantic Salmon Symposium (1999); and Derek Mills, ed., *The Ocean Life of Atlantic Salmon: Environmental and Biological Factors Influencing Survival* (London: Fishing News Books, 2000). On the New England fishery, see the articles collected in "The History, Status, and Future of the New England Offshore Fishery," *Northeastern Naturalist* 7 (2000).

36 See the North Atlantic Salmon Fund Progress Report, June 1996. See also the National Fish and Wildlife Foundation report, "Ending the Greenland Atlantic Salmon Fishery: The U.S. Position and the Prospects for Funding" (November 15, 1994). There are also threats beyond the reach of any political jurisdiction, such as climate changes and increases in sea-surface temperatures.

37 See the North Atlantic Salmon Fund Progress Report, June 1996. See also the National Fish and Wildlife Foundation report, "Ending the Greenland Atlantic Salmon Fishery: The

U.S. Position and the Prospects for Funding" (November 15, 1994). There are also threats beyond the reach of any political jurisdiction, such as climate changes and increases in sea-surface temperatures.

38 The members of the Maine Salmon Rescue Coalition include Maine State Chamber of Commerce, Maine Wild Blueberry Commission, Maine Forest Products Council, Bangor Hydro-Electric Company, Penobscot Hydro LLC, Maine Pulp & Paper Association, Maine Aquaculture Association, and FPL Energy, LLC. See "Comments of the State of Maine in Opposition to the Proposed Endangered Status For a Distinct Population Segment (DPS) of Atlantic Salmon in the Gulf of Maine." See also Irv Kornfield, John Bailey, Ken Beland, and Chuck Ritzi, "Report of the Salmon Genetics Committee" (December 19, 1995), reproduced as Appendix 2 in "Comment of the Maine Salmon Rescue Coalition." On the lowering of fitness, see Philip W. Hedrick and Steven T. Kalinowski, "Inbreeding Oppression in Conservation Biology," *Annual Review of Ecological Systems* 31:139–162 (2000). The authors of "Comments of the Maine Salmon Rescue Coalition" argue that the Services fail to analyze the role of genetic drift and ignore the "Report of the Salmon Genetics Committee." They also argue that the Services ignore research that indicates fairly high straying rates of salmon. The potential for genetic drift, however, found in small isolated populations, is mitigated by high stray rates in migratory animals such as Atlantic salmon that introduces new genes into a population.

39 Babbitt's comments, transcribed from a tape recording, are reproduced in the "Comments of the Maine Salmon Rescue Coalition," Appendix 20.

40 The Governor of Maine directed various state agencies to implement the Conservation Plan, under the leadership of the Land and Water Resources Council, to be chaired by the director of the State Planning Office. Members of the Land and Water Resources Council include commissioners of the following departments: Environmental Protection, Conservation, Marine Resources, Agriculture, Transportation, Human Services, Economic and Community Development, and Inland Fisheries and Wildlife. The Council created an Atlantic Salmon Committee that, in addition to department commissioners, included the Chair of the Atlantic Salmon Authority and representatives from each Watershed Council. See the Maine Atlantic Salmon Task Force, "Atlantic Salmon Conservation Plan for Seven Maine Rivers," (March 1997). See also the 1999 and 2000 Annual Progress Reports, "Atlantic Salmon Conservation Plan for Seven Maine Rivers," Maine Atlantic Salmon Commission, at www.state.me.us/asaRaymond J. O'Connor, Ray B. Owen, and Judith Rhymer present a rather optimistic assessment of the strengths of the state plan, in "Best Practices in Endangered Species Recovery Planning: Lessons for the Conservation of Maine's Atlantic Salmon," *Maine Policy Review* 9:72.

41 See U.S. District Court for the District of Columbia, Defenders of Wildlife, et al., v. Bruce Babbitt, et al. (Civil Case no. 99-CV-00206); and Trout Unlimited, et al., v. Bruce Babbitt, et al. (Civil Case no. 99-CV-02143). The Services declarations to the court can be found as part of the documents in the "Comments of the Maine Salmon Rescue Coalition," Appendices 12 and 14.

42 The 1999 "Biological Report on the Status of Atlantic Salmon" can be found at http://news.fws.gov/salmon/asalmon.html.

43 The NAS committee met on June 12–13, 2001 and again on September 20–22, 2001 in Bangor, Maine. King's speech, "Restoration Not Regulation," (December 2, 1999), can be found at www.state.me.us/govemor/policy/salmonspeech/salmon2_text.htm.

44 James T. Carlton, Jonathan B. Geller, Marjorie L. Recha-Kudla, and Elliot A. Norse, "Historical Extinctions in the Sea," *Annual Review of Ecological Systems* 30:515–538 (1999).

45 Local extinction in coastal ecosystems is primarily caused by overfishing, and other human disturbances such as pollution, climate change, degraded water quality, etc., come later. The ecological effects of the nineteenth-century overfishing of Atlantic salmon have not been systematically studied. See Jeremy B.C. Jackson, et al., "Historical Overfishing and the Recent Collapse of Coastal Ecosystems," *Science* 293(5530):629–638 (July 27, 2001). See also David K. Cairns, "An Evaluation of Possible Causes of the Decline in Pre-fishery Abundance of North American Atlantic Salmon," *Canadian Technical Report of Fisheries and Aquatic Sciences* 2358:i–vii, 1–67 (April 2001).

46 These comments can be found at www.senate.gov/~snowe/salmontest.htm and also at www.senate.gove/~collins/991228.htm. Collins might have also said that inadequate science, flawed logic, and bad policy combined to produce salmon populations declines in the first place.

47 "Final Endangered Status," 65 *Federal Register* 69459 (November 17, 2000). Some people fear the river-specific hatchery fish will result in increased genetic drift within particular rivers resulting in decreased fitness, with the potential of causing the final extirpation of Atlantic salmon in Maine. See "Comments of the Maine Salmon Rescue Coalition."

48 "Final Endangered Status," 65 *Federal Register* 69459.

49 *Endangered and Threatened Species Recovery Program*, prepared by the U.S. Department of Interior, U.S. Fish, and Wildlife Service (Washington, DC: Government Printing Office, 1990), v. The George W. Bush administration has attempted to forestall further listings by curtailing the funds available to the Fish and Wildlife Service for the study of new listings.

50 "Genetics Status of Atlantic Salmon in Maine," Interim Report from the Committee on Atlantic Salmon in Maine, prepublication copy. Available at http://nap.edu/openbook. The final interim report was made available March 2002 (Washington, DC: National Academy Press).

51 "Comments of the Maine Salmon Rescue Coalition," 10 ff.

52 "Comments of the State of Maine," 4.

53 "Comments of the Maine Salmon Rescue Coalition," 3. One hopes that such comments are not self-fulfilling prophesies, given the fact of the endangered species listing of Atlantic salmon.

54 "Comments of the Maine Salmon Rescue Coalition," 3.

55 I am paraphrasing Richard White, whose observations about science and policy in his study of Pacific salmon and the industrial development of the Columbia River parallel those developed here. *The Organic Machine: The Remaking of the Columbia River* (Hill and Wang, 1995), 106.

56 Susan Young, "Salmon Hatchery Dedicated," *Bangor Daily News* (September 30, 2002).

57 "Endangered and Threatened Species; Designation of Critical Habitat for Atlantic Salmon (Salmo Salar) Gulf of Maine Distinct Population Segment," 74 Federal Register 29299 (June 19, 2009).

58 "Annual Report of the U.S. Atlantic Salmon Assessment Committee. Report Number 31 – 2018 Activities" (Portland, ME, 2019), 15. www.nefsc.noaa.gov/USASAC/Reports/USASAC2019-Report-31-2018-Activities.pdf.

59 "Whole Oceans Open Bucksport Facility," *Bangor Daily News* (August 10, 2020).

4

COUNT EVERY FISH

Nonmarket Fishing Economies on the Yukon River

I first heard the tale of the pickup truck in 2010. Someone on the middle portion of the Yukon River had sold enough Chinook salmon, caught for human subsistence, to buy a truck worth $15,000. Of this wide-ranging rumor, details were murky. There could have been one fisherman or perhaps an extended family, whose members combined the proceeds of their catch to purchase the pickup, possibly as a gift for a mother-in-law. The story was told to promote an argument of foul deeds being committed. Fish caught for food was being turned into a commodity for profit. As the story circulated, the value of the truck increased to $20,000. In the context of scarcity and food security, the profit motive had undermined traditional values of subsisting on and sharing fish caught from the Yukon River. The last time I heard the story in 2012, the price of the truck had risen to $40,000. That represented an amount of fish that should have fed families, not purchased someone's 4×4.[1]

The story about salmon providing pickups became prominent in 2010 and continued for a few years thereafter. The timing makes sense in part because the prior year's fishing was unusual. In 2009, some 879,185 salmon of all species had been harvested in the subsistence fishery in Alaska. This was the lowest subsistence harvest of salmon in over 16 years of record-keeping. That year 33,932 Chinook salmon were harvested for subsistence and personal use on the Yukon River, well below the ten-year average of 55,510.[2]

As a protective measure, the first pulse of Chinook on the Yukon River was closed entirely to fishing in 2009, as managers attempted to allow more fish to cross into Canada to spawn to make up for years of declining runs. Subsistence, commercial, and sport fishing were all precluded from targeting the first pulse of Chinook, as a rolling closure followed the fish upriver into Canada (Figure 4.1).

The Yukon River commercial harvest in 2009 totaled 316 Chinook. Virtually the entirety came from the Lower Yukon River outside of the rolling closure and were incidentally caught during a fishery effort targeting other species. The harvest

DOI: 10.4324/9781003297444-6

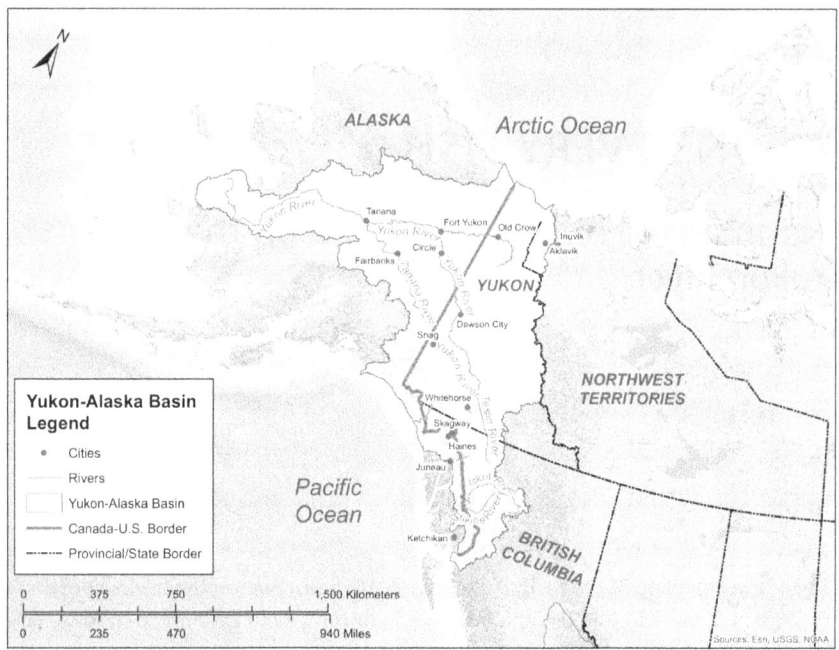

FIGURE 4.1 Map of the Yukon River Drainage

Source: International Joint Commission, https://ijc.org/en/watersheds/alaska-yukon.

indicated a 99 percent decrease from the ten-year (1999–2008) average of 35,000 commercially harvested Chinook. The other salmon species commercially harvested that year included 8,000 Coho and 195,000 chum salmon.

Two buyer-processors operated in the Lower Yukon Area and two in the Upper Yukon Area, which paid $0.50 per pound for summer chum in the lower river and $0.24 per pound for summer chum in the upper river. The average income for fishers in the Lower Yukon Area was $1,425 in 2009, compared to $1,857 for Upper Yukon Area fishers. In this context, the story of a $15,000–$40,000 truck bought from the sale of subsistence-caught Chinook generated considerable anger among fishers from the lower river. Some lower Yukon River fishers perceived sales of subsistence-caught Chinook from the middle and upper reaches of the Yukon in direct competition with their commercial interests, and they sought to curtail the practice.[3]

People who live along the Yukon River eat subsistence-caught fish, but they also sell it outside of the market economy, which has caused no end of confusion and debate, and considerable anxiety—for both natives and non-natives. The confusion stems from the coexistence of different kinds of economies. The debate emerges between those who recognize only a Western market economy as legitimate, and those who continue to practice traditional economic exchanges for which supply and demand, competition, cost/benefit calculations, and other market forces are irrelevant. Both sides fail to understand that modern economies, especially at the peripheries of the capitalist system, are never monolithic; they are always mixed. The anxiety derives from the fact that certain fish stocks are in sharp decline, with Yukon River Chinook salmon at the top of the list. The consequences of the decline are cultural, economic, and environmental—categories that remain difficult to tease apart with any precision, providing additional layers of confusion, debate, and anxiety. In such contexts, tales of trucks in bush Alaska proliferated. Urban legends are anemic by comparison.

In microcosm, "customary trade"—the exchange of cash for subsistence-caught foods—illuminates the complexity of human-environment relations and provides yet another example of how deeply entangled human and natural systems have become. Neither an analysis of natural systems nor one of social systems is sufficient to illustrate the entanglement. Natural and social environments intertwine at multiple scales and with multiple effects.

We live in a world of coupled human and natural systems, not two separate systems, with all the ambiguity and uncertainty that implies. The categories of nature and culture, of spawning fish and policy development, of ecosystem and market, cannot be neatly cleaved apart. The metaphors of ecological balance and harmony at one extreme, or of environmental destruction and war at the other, do not easily apply to coupled human and natural systems. This is certainly the case for Yukon River Chinook salmon, and for the small-scale nonmarket exchanges of cash for fish that have come under scrutiny by those who wish to control or eradicate the practice, substituting the more tractable rationality of Western markets.

By commercial standards, the subsistence fishery targeting Chinook salmon on the Yukon River and its tributaries is minor. In discussions of customary trade, the scale of harvest should be kept forefront. Measured in pounds, subsistence harvests of fish and game comprise about 1.1 percent of total harvests of wild resources in Alaska. Commercial fisheries account for 98.3 percent of wild resource harvest, while sport hunting and fishing account for 0.6 percent. The scale of human population should also be kept forefront: in Alaska, about 12,000 people live in small villages along the Yukon and its tributaries. Another 100,000 live along the Tanana River, the second largest drainage of the Yukon after the Porcupine River. Most of these people live in Fairbanks North Star Borough, the second largest population center in Alaska after Anchorage. Under federal rules, those who live in Fairbanks are ineligible for the federal subsistence priority but can hunt and fish under state rules.[4]

Given the relative scales of harvest, and the small human population along the Yukon River, why did fisheries managers and state and federal bureaucrats become concerned with minor exchanges of cash for fish? What was at stake? What bureaucratic effort was at play? Who, in the event, was "authorized" to insist on changes to local cultural practices? To what end?

I think the answer to these questions partially lies in the broader context of neoliberal economic restructuring, which tends to subordinate social and environmental policies to economic policies. Neoliberal economic policies, however, are ill-equipped to accommodate cultural values and practices that do not conform to the logic of markets, especially in situations of natural resource governance. I'll pursue here one general line of argument, knowing the diversity of contexts presses against that argument.

I argue that the logic of the market is used in Alaska to undermine practices that do not adhere to that logic. This happens in two ways. First, by characterizing nonmarket economies as market economies, fisheries managers, and bureaucratic decision-makers use the logic of the market to demonstrate the purported incoherence of practices that stand outside of its reach. Second, by using economics rather than anthropology as foundational, managers and decision-makers speak the language of numbers and data sets, statute and regulation, costs and benefits, rather than the language of culture and local meanings; they thereby diminish the significance of cultural practices that avoid quantification and bureaucratic control. This is evident in Chapter 5 and is equally evident here. Not everything is monetizable. Not all values relate to money.[5]

This chapter describes a relatively minor, mostly native nonmarket economy founded on subsistence-caught Chinook salmon that became the object of controversy and increasing regulatory pressure. It also describes the cultural consequences of that pressure for native and other rural peoples along the Yukon River, Alaska. I show that customary trade practices in Alaska are of bureaucratic concern because (1) they are elusive to state observation, (2) they stand outside of easily quantifiable techniques of data collection, and (3) they do not fit Western economic rationality. In addition, people on the Yukon River resist participating in the quantification of their customary trade practices. As Yukon River Chinook runs declined, state and federal resource managers started to regard customary trade as inimical to managing Chinook fisheries. Managers became worried they were failing both the fish and the humans who relied on them, when they failed to count every fish—and fish exchanged for cash were especially hard to track.

To get to all of this requires a discussion of interconnected social and environmental contexts in which to situate small-scale exchanges of cash for subsistence-caught fish. These contexts include legal definitions and related policy issues, state and federal management regimes, worldwide markets, differences between market and nonmarket economies, incompatible cultural values, and bureaucratic concerns over the health safety of traditionally processed fish. Courts weigh in. Biologists and anthropologists wave their arms in alarm. Botulism makes an unexpected appearance. Withal, rural peoples continue their traditional practices. At every turn, various bureaucratic efforts appear halting, ad hoc, and ill-informed—or at best, partially informed—even as individual fisheries managers themselves were well-intentioned. Yukon River Chinook runs were in perilous decline, and fisheries managers did not understand why.

Subsistence Laws and Regulations

Federal and state laws in Alaska governing the sale of fish caught for subsistence are contradictory and, in any case, largely ignored in practice. Prominent among the contradictory laws are the 1980 Alaska National Interest Lands Conservation

Act (ANILCA) and its implementing regulations, which allow for the sale of subsistence-caught fish, and State of Alaska regulations, which do not, with two minor exceptions.[6]

ANILCA designated 217 million acres of federal lands in Alaska as "conservation system units." These units included vast parks, preserves, and refuges, totaling some 60 percent of the state.

Title VIII of ANILCA uniquely provided rural peoples the opportunity to continue subsistence hunting, fishing, and gathering on federal public lands. Congress wrote that

> the continuation of the opportunity for subsistence uses by rural residents of Alaska, including both Natives and non-Natives, on the public lands and by Alaska Natives on Native lands is essential to Native physical, economic, traditional, and cultural existence and to non-Native physical, economic, traditional, and social existence.
>
> *Title VIII Sec. 801 (1)*

Congress did not explain why it included the term "cultural" in reference to native existence but not in reference to non-native existence. Perhaps Congress recognized a unique native cultural relationship to federal public lands which more recent immigrants to Alaska did not have.

This recognition of native culture, and the development of Title VIII in ANILCA, fulfilled a congressional promise from a prior act, the 1971 Alaska Native Claims Settlement Act (ANCSA), which extinguished aboriginal title, abolished all native hunting and fishing rights, and substituted corporation for tribal ownership to 45 million acres of land—in effect, imposing a corporate, profit-oriented structure onto native-owned lands. In ANCSA, Congress left unresolved hunting, fishing, and other subsistence activities of native peoples, but anticipated addressing the issue at a later date. In a political compromise, Congress eventually provided a subsistence priority to all "rural" residents, thus avoiding a politically difficult native priority for hunting and fishing on federal lands.[7]

Section 803 of ANILCA defined "subsistence uses" to mean

> The customary and traditional uses by rural Alaska residents of wild, renewable resources for direct, personal or family consumption as food, shelter, fuel, clothing, tools, or transportation; for the making and selling of handicraft articles out of nonedible byproducts of fish and wildlife resources taken for personal or family consumption; for barter, or sharing for personal or family consumption; and for customary trade.

Congress has defined "barter" but not "customary trade." Barter

> means the exchange of fish or wildlife or their parts, taken for subsistence uses—(A) for other fish or game or their parts; or (B) for other food or for

nonedible items other than money if the exchange is of a limited and non-commercial nature.

Note that barter and sharing were limited to "personal or family consumption." Leaving aside the question of why Congress felt it necessary to tell rural citizens of Alaska they were permitted to share with one another, what was customary trade? On this question the statute is silent. However, one significant difference is that, unlike barter and sharing, Congress did not link customary trade to personal or family consumption; indeed, customary trade was not linked to any purpose, perhaps because, for Congress, the purpose of cash exchanges was self-evident.[8]

A report from the Senate Committee on Energy and Natural Resources, which comments on ANILCA, is frequently cited as the basis for customary trade regulations implementing Title VIII. "The Committee does not intend that 'customary trade' be construed to permit the establishment of significant commercial enterprises under the guise of 'subsistence uses.'"[9] Commercial enterprises were out, noncommercial enterprises were in. What, then, was customary trade? Was it simply, to borrow part of the definition of barter, the exchange of cash for "fish or wildlife or their parts, taken for subsistence uses … if the exchange is of a limited and noncommercial nature"? If so, what would the exchange of cash mean in a nonmarket economy?

The ethnographic literature on this topic indicates a surprising range of meanings associated with cash in a variety of cultural and economic settings. The use, importance, and meaning of cash in nonmarket economies are far from self-evident. Moreover, when did customary trade end and a significant commercial enterprise begin? Who would make these determinations, based on what criteria?[10]

Congress anticipated the State of Alaska would administer hunting and fishing regulations on federal lands and provide a subsistence priority to "rural" residents. Prior to the passage of ANILCA, the state had passed its own subsistence regulations in 1978; much of the state's statutory language was in fact adopted into the federal statute. After the passage of ANILCA, the state regulated subsistence activities throughout Alaska for nearly a decade. A series of court rulings, however, altered the administrative landscape and produced a rather odd state-federal dual-management system, which was to be temporary, pending an amendment to Alaska's state constitution. That amendment hinged on the definition of "rural." Cash exchanges were also implicated, even if poorly understood.

In 1988, the Ninth Circuit Court of Appeals, deciding *Kenaitze Indian Tribe v. State of Alaska*, ruled that that state's definition of "rural" did not comport with the word's meaning in ANILCA. "We travel to the northern reaches of our circuit," the court wrote,

> to resolve a dispute implicating two recurring Alaskan motifs: on the one hand, the clash between traditional and modern ways of life; on the other, fish. The Kenaitze Indian Tribe claims that the state of Alaska is attempting to

evade federal legislation creating a priority for subsistence fishing by residents of rural areas. The controversy turns on the meaning of the word "rural" as used in the Alaska National Interest Lands Conservation Act.

The appellate court noted rapid changes to local area economies and practices, which occasioned the present controversy:

> The Kenaitze, a tribe numbering approximately four hundred, have lived on the Kenai Peninsula, in southern Alaska, for hundreds of years. For most of their history, the Kenaitze have pursued a way of life dominated by subsistence fishing and hunting. In recent years, however, the area's proximity to Anchorage has made the Kenai Peninsula a center of commercial and sport fishing, and has transformed the Peninsula's economy to one based primarily on work for cash. Subsistence fishing has been crowded out by commercial harvesting and by sport fishing, the latter pursued with all the zeal of a Crusade.

The state statute—first through regulations promulgated by the Boards of Fish and Game, then through legislative redefinition—defined "rural" based on ten criteria. The result was a characterization of "rural," as "a community or area of the state in which the noncommercial, customary, and traditional use of fish or game for personal or family consumption is a principal characteristic of the economy of the community or area."[11] Where market-based economies prevailed, there "rural" Alaska communities did not exist. This was, in its way, a negative definition: it negated communities whose members relied on subsistence if those communities had been engulfed by the Western economy.

The court found this definition entirely inadequate, and contrary to ANILCA. "The state has selected an unusual definition of the term rural," the court noted, "a definition that excludes most areas normally understood to be covered by the term." The court went on to explain:

> As Alaska defines the word, an area is rural only if its economy is dominated by subsistence fishing and hunting; it excludes areas characterized primarily by a cash economy, even though a substantial portion of the residents may engage in subsistence activities. The state's definition would exclude practically all areas of the United States that we think of as rural, including virtually the entirety of such farming and ranching states as Iowa and Wyoming.

The court further noted that "the state's contorted definition of rural would materially change the sweep of the statute, second-guessing the congressional policy judgment embodied in ANILCA. This we may not do." Alaska was not "required to regulate pursuant to ANILCA," the court acknowledge; but if it chose to do so "it is bound to implement the statute as Congress passed it, not as some of its citizens would prefer that it had been passed." The court summarized its opinion in the following strongly worded paragraph:

> We end as we began. This is a case involving a clash of lifestyles and a dispute over who gets to fish. Congress, using clear language, has resolved this dispute in favor of the Kenaitze who choose to pursue a traditional subsistence way of life by giving them priority in federal waters. The state has attempted to take away what Congress has given, adopting a creative redefinition of the word rural, a redefinition whose transparent purpose is to protect commercial and sport fishing interests.[12]

This case of "who gets to fish," is also indirectly a case of "who gets to exchange cash for fish." Congress gave priority to rural residents over commercial and sport fishers—including those sport fishers with the zeal of a Crusade—and that priority included customary trade.

State regulation of subsistence practices, however, became even more problematic in 1989 when the State Supreme Court found that the state's constitution precluded a priority for "rural" over "nonrural" residents and thus the state could not comply with the mandates of ANILCA even if it managed to develop a suitable definition of "rural."[13] All residents must be treated equally under Alaska's constitution, the court found, with none receiving benefits not accorded others. This ruling caused a particular, long-lasting problem. The state could not implement what was constitutionally prohibited. Yet Title VIII of ANILCA needed to be implemented, despite the state's constitutional inability to do so.

The Federal Subsistence Program was thus established with the publication of temporary subsistence management regulations in 1990.[14] The widespread presumption was the state would amend its constitution and fish and wildlife administration would return fully to state control. Despite a great deal of legislative effort over several years, the state constitution remained as it was. In the wake of the state's failure to amend its constitution, an awkward and contentious system of dual federal and state management evolved. Rural people were caught between adversarial bureaucracies, as they continued to practice their own subsistence lifestyles.

In the uncertain political and legal contexts of state and federal control over resources, waterways, and subsistence uses, one of the many issues the Federal Subsistence Program attempted to resolve, with limited success, was customary trade. The federal program began to explore this issue in 2000. It quickly became apparent that customary trade was implicated in a range of other issues, including changing environments, state-federal animosity, declining Chinook runs, U.S.-Canada agreements, fish markets, health concerns, and differences between native and Western economic practices. Without declining Chinook salmon populations, customary trade would have remained an obscure and tolerated practice. As it turned out, however, Chinook population declines, and questionable management decisions, brought customary trade into greater public focus.

Let me make a small aside: Neither the courts nor the federal program have examined the contradictions between the profit-oriented corporate structures established by ANCSA, and the nonmarket customary trade practices allowed under ANILCA. David Harvey notes that "a signal feature of the neoliberal project," is the "corporatization, commodification, and privatization of hitherto public assets." ANCSA fits securely within this project, with its corporatization of tribal lands. ANILCA, by contrast, with its emphasis on cultural continuity, not profit maximization, provides a statutory limit to the neoliberal project. With these two acts, Congress provided a landscape of confusion, made more confused with regulations imposed by the State of Alaska. This was the world I has hired to help navigate.[15]

Managing Environments for Subsistence

The Yukon River originates in British Columbia, Canada, and flows 2,300 miles north and then west into Alaska before it reaches the Bering Sea. The drainage for the river is immense and encompasses over 330,000 square miles. Chinook spawn in many of its tributaries. As migratory animals, Yukon River Chinook salmon (*Oncorhynchus tshawytscha*) live in a variety of complex environments, from river to estuary to open ocean and back again to spawn and then die in their natal stream. They may become sexually mature after one winter at sea—these are usually males—after which they return to spawn, or they may become sexually mature after two to seven years at sea. In any given year, a mix of age classes returns to spawn, an evolutionarily stable strategy that ensures a single spawning year will not die out as a result of some catastrophic environmental event. Of the five species of Pacific salmon, Chinook are the largest and may exceed 50 pounds. In the early twentieth century, Chinook caught from the Yukon River often weighed well over 70 pounds. At mid-century, the largest commercially caught Chinook on record, from southeast Alaska, weighed 126 pounds.[16]

Chinook salmon are prized because of their high oil content, exceptionally large size, and what were once regular rates of return. They are prized because humans make use of them. At one time, such use was proportional to the numbers of returning salmon. With the turn of the twentieth century, however, as Europeans established themselves along the Yukon River, more and more salmon were harvested—to feed people, to feed commercial markets, and to feed dogs (whose numbers increased in response to the transportation needs of the fur and gold mining industries).[17]

By the beginning of the twenty-first century, Chinook began to decline rapidly and occasioned heightened concern. In addition to the decline in absolute numbers, the size of the fish was declining as well. In 2010, the average Chinook salmon caught at Rampart Rapids, 763 miles upriver, weighed just under 11 pounds. Moreover, the ratio of males to females appeared to be changing dramatically, with proportionally fewer females returning to spawn. Dan O'Neill, in his assessment of managerial failure, puts the matter thus:

To a fisheries biologist, these are classic signs of a fish stock in peril. Except, apparently, if the biologist happens to be managing this fishery for the Alaska Department of Fish and Game or the US Fish and Wildlife Service. In any case, the slender thread connecting the people along the Yukon to eight thousand years of traditional fishing appears ready to snap.

In case the reader misses his point, O'Neill reiterates it in stronger terms, highlighting what he sees as management's concern for commercial interests over the long-term value of salmon:

> The shrinking fish problem … was abetted by the managers of the two agencies—the Alaska Department of Fish and Game (ADF&G) and the US Fish and Wildlife Service—that have jurisdiction over the Yukon River salmon runs. There were some excellent field biologists working on the river, but their superiors, the fishery managers, seemed oblivious to all the warning signs of a fish stock in serious decline—perhaps on the brink of collapse—while obstinately accommodating the downriver commercial fishing interests. Reciprocally, managers' policies that favored the commercial industry necessarily short-change the upriver subsistence users.[18]

Chinook salmon live in rivers, estuaries, and the open ocean, but they also live in human-imposed districts, some of which were put in place to facilitate natural resource management. The Yukon River is managed as a single river, but to do so requires segmenting it into manageable units and delimiting its salmon populations for commercial, sport, and subsistence purposes. In addition, the United States is obligated by treaty to allow the passage of 42,500–55,000 Chinook into Canada but frequently fails to meet that obligation. In 2014, however, over 64,000 Chinook passed the border into Canada, marking a change in the pattern. About 50 percent of Yukon River Chinook originates in Canada. Failure to ensure sufficient numbers escape the U.S. gauntlet into Canada simply means diminished and diminishing future returns.[19]

For management purposes on the U.S. side, the Yukon River is divided into seven districts, one of which is coastal. These are further subdivided into ten subdistricts. For subsistence, about half of the Yukon River drainage falls under state jurisdiction and half under federal jurisdiction. Different gear types (e.g., drift gillnets, set gillnets, fish wheels), a limited and variable duration of the subsistence fishery based on in-season estimates of returning Chinook, different reporting requirements of subsistence harvests, and whether one has rural status or not, are used to control subsistence fishing opportunities, depending upon the district. Otherwise, within the federal program, there are no limits to subsistence fishing for Chinook, absent outright closures.

Although the Yukon River is managed as a single system, it flows through and adjacent to diverse federal, state, native corporation, and other private and municipal lands. The federal units—administered by the Bureau of Land Management, the Fish and Wildlife Service, and the National Park Service—all have different mandates for conservation and different strictures on human use. Residents often complain that the resulting patchwork of ownership and rules makes following regulations unnecessarily complicated. In some instances, rural residents fishing in waters under federal jurisdiction may be subject to state rules when standing on state or private lands, but if they take one step into a federally controlled waterway, then they are subject to federal rules. Where the fish was hauled ashore appears more important than where the fish was swimming. The state also asserts that "While standing on state and private lands (including state-owned submerged lands and shorelands), persons must comply with state laws and regulations and cannot sell subsistence-caught fish." Where the fish was traded for cash appears more important than where the fish was caught, federal jurisdiction notwithstanding. These interpretations of state and federal authority seem, as some have commented, more a petulant expression of state power than a rational articulation of resource management.[20]

As required by its constitution, the State of Alaska manages according to "the sustained yield principle" in order to develop "natural resources to maximize benefits for Alaskans." The relevant section of the state constitution reads as follows: "Fish, forests, wildlife, grasslands, and all other replenishable resources belonging to the State shall be utilized, developed, and maintained on the sustained yield principle, subject to preferences among beneficial uses." One such beneficial use, indeed the highest preferential use under state and federal law, is subsistence. In other words, commercial interests, as promoted by neoliberal policies, should be given a lower priority; but in many instances, commercial interests are allowed a prominence that, by law, they should not enjoy.

"Surplus" Fish

To manage Chinook salmon for subsistence requires five distinct sets of numbers. The first is the number of Chinook salmon in the river system; the second is the number of salmon allowed to follow their migratory path unimpeded to spawning grounds, called "escapement," which is to ensure the persistence of the species; the third is the number harvested for commercial purposes; the fourth is the number harvested for sport; and the fifth is the number harvested for subsistence. The last three numbers are considered "surplus" fish to be targeted for human interception and use.

Under this model, all surplus fish should be caught to maximize human benefits and to avoid what the Alaska Department of Fish and Game calls "over escapement." The claim is that over escapement—that is, more than some optimal number of fish on the spawning grounds—produces fewer fish.[21] While this may be true under certain limited conditions, it ignores the benefits to the ecosystem provided by large numbers of returning salmon. Returning salmon are food sources for a diversity of animals, from bears to eagles, and as they die and decompose salmon contribute a huge nutrient load to the river ecology. Returning salmon transport energy from marine to freshwater systems. Precluding that energy from entering the freshwater system, and instead diverting it to human use, may eventually result in further degrading the very system humans are trying to maintain. Maximizing short-term benefits for Alaskans in the form of surplus fish may minimize long-term benefits for the larger ecosystem upon which both fish and Alaskans depend.[22]

Of the surplus Chinook, the Alaska Board of Fisheries has allocated 44,500–66,704 fish as "amounts necessary for subsistence" for area residents. For the 12,000 persons who live in the Yukon River drainage, this means 3.7–5.6 fish each. If these fish weigh over 50 pounds apiece, then the subsistence harvest is substantial, ranging from 185 to 280 pounds per person. If these fish average 11 pounds apiece, then the "amounts necessary for subsistence" range from 40.7 to 61.6 pounds per person.

Somewhere within these amounts are the fish exchanged for cash as customary trade, a practice that is allowed under the federal rules as a legitimate subsistence use but precluded under state rules. But the extent of customary trade remains obscure for any number of reasons, including the methods of estimating both Chinook numbers and subsistence harvests. In addition, the federal program does not set "amounts necessary for subsistence," presumably because those amounts are best left to the local peoples to determine, not a citizen board appointed by the governor. As legal scholars David Case and David Voluck point out, "The boards … make wildlife management policies in splendid isolation from the rural (predominately Native) populations, which are most heavily affected by these policies."[23] The boards may also make policies isolated from the long-term ecosystem effects such policies produce.

In order to determine surplus, the number of migrating Chinook must first be estimated. Managers generate estimates when the fish pass a sonar at the village of Pilot Station, located 120 miles from the mouth of the Yukon, and again as they pass a sonar at the village of Eagle near the U.S./Canadian border. Three additional sonar stations have been placed on tributaries. Sonar, however, cannot clearly distinguish species of fish. To ascertain the relative abundance of all species of salmon passing through each area, the Alaska Department of Fish and Game captures salmon by drift gillnet at five test fisheries at different locations on the Yukon River, the uppermost below Kaltag 450 miles from the mouth of the river. Chinook have also been counted at five weirs on tributaries of the Yukon, from five towers and nine fish wheels, and by aerial surveys.

The effort is expensive, logistically challenging, and not entirely reliable. It is used for both in-season management and postseason assessment. However, the Alaska Department of Fish and Game does not generate formal projections of salmon runs in most of the Arctic, Yukon, and Kuskokwim areas and instead offers annual forecasts that "are qualitative in nature because of a lack of information with which to develop more rigorous forecasts." A lack of information means that overall populations are unknown, which then calls into question escapement goals and allocations of surplus. Overestimating population size results in over-allocations of surplus, potentially accelerating rates of declines. By contrast, underestimating population size results in under-allocations of surplus, potentially causing subsistence fishers to go hungry.[24]

The fact of sharp declines of Yukon River Chinook salmon may be taken as evidence of the failure to adequately manage the fish. Indeed, such declines may be the direct result of management practices. Or declines may be evidence of historically occurring fluctuations of population size that resource managers do not understand and cannot effectively manage. As the Alaska Department of Fish and Game concludes, "Yukon River Chinook salmon are currently considered to be in a period of low productivity," effectively shifting blame to the natural world.[25] In any event, population fluctuations fail to fit sustained yield principles geared toward maximizing short-term human benefit. Although the natural world is more unpredictable than the Alaska constitutional provision allows, management by sustained yield attempts to remove elements of unpredictability—apparently in this instance without much success. It may be that sustained yield practices, inaccurate population estimates, and over-allocations based on inaccurate models of human use generate what Dean Bavington, in his study of the Newfoundland cod fishery collapse, calls "managed annihilation."[26] Yukon River Chinook salmon may suffer a similar fate.

Once a total population size is estimated, and escapement is subtracted, the "surplus" remains. The subsistence harvest is estimated through in-person and phone interviews with Yukon area residents who volunteer to participate in post-season household surveys. Data from interviews is then combined with information from harvest calendars upon which people may record their catch. Surveys and interviews are conducted by the Alaska Department of Fish and Game, Division of Subsistence. As with counting fish, this effort is also expensive, logistically challenging, and not altogether reliable. The voluntary harvest calendars, for example, have about a 15 percent rate of return. Such survey and calendar data are useful only for postseason assessment, not in-season management. And it should be pointed out that collecting data for sustained yield and allocation models precludes other ways of evaluating the significance of natural resources. These models predetermine the kinds of data researchers collect and limit the resulting interpretations, vastly oversimplifying nature and culture and underplaying the dynamic of their interaction.[27]

The Bering Sea: Managing Chinook Bycatch

The open ocean fishery further complicates Chinook management, in part because neoliberal policies often conflict with other management concerns, and in part because of the changing nature of ocean environments. The Bering Sea, subject to a 200-mile U.S. Exclusive Economic Zone, is divided into two areas, the Bering Sea subarea and the Aleutian Islands subarea. The Aleutian Island subarea is further divided into three districts. The Bering Sea subarea pollock (*Theragra chalcogramma*) fishery, which has had a substantial impact on Chinook salmon, is required to adhere to a hard cap of 60,000 Chinook caught as bycatch, a cap the industry views as an intrusion into its market viability. This cap, however, requires voluntary participation and is distributed across four sectors of the industry: the trawl/processor sector, the mothership sector, the inshore sector, and the Community Development Quota sector. In the event the cap is exceeded, stronger prohibitions may be imposed, including the closure of the pollock fishery.[28]

Subsistence and commercial fishers on the Yukon River frequently cite the pollock industry as the major cause of Chinook declines and have expressed frustration with the inability of the federal government to find an adequate solution. The bycatch cap, however, applied in 2012, responds to concerns over Chinook declines, but it also tries to balance the pollock fishery optimal yield against the value of Chinook salmon. A lack of complete information hampers this balancing act. The National Oceanic and Atmospheric Administration (NOAA) and the National Marine Fisheries Service (NMFS), which oversee the pollock fishery, attempt to

> minimize bycatch to the extent practicable given the tools currently available to the fleet, the derby-style prosecution of the fishery, the uncertainty about whether the bycatch has adverse effects on any particular Chinook salmon stocks, and the need to ensure that the pollock fishery contributes to the achievement of optimum yield in the groundfish fishery.

NOAA and NMFS recognize that diverse user groups have a stake in Chinook salmon, including "commercial, recreational, and cultural user groups." However, "With the information that is currently available, neither the total 'cost' of Chinook salmon PSC [prohibited species catch] nor the total 'value' of Chinook salmon savings can be estimated for the various user groups." NOAA and NMFS are left with best guesses, constrained by the fact that Chinook salmon are "fully allocated," which means there is little room for error.[29]

Other significant factors in Chinook declines include climate change, ocean acidification, fisheries-induced evolution, and long-term fluctuations of fish populations—long-term referring to a few centuries, not a few decades. The scientific understanding of long-term fluctuations of fish populations is surprisingly thin. These factors remain beyond the ability of Yukon River managers to manage, especially when combined with the increasing intrusion of the market economy into remote regions of the Yukon River. The apparent historical lesson, which hardly needs another example, is this: natural and economic systems do not neatly align.[30]

Markets and Subsistence

The market economy intrudes into discussions of sales of subsistence-caught Chinook salmon in two prominent arenas. The first and most prominent is the intercept fishing industry targeting pollock in the Bering Sea, which catches substantial numbers of Chinook salmon destined to spawn in the Yukon and its tributaries. The second is the Western Alaska Community Development Quota Program (CDQ), begun in 1992 under the Magnuson-Stevens Fishery Conservation and Management Act, which allocates a portion of the Bering Sea/Aleutian Islands fisheries catch for job creation and infrastructural development in 65 native communities in western Alaska.[31]

The bycatch of Chinook salmon in the Bering Sea has declined in recent years, both because of changed industrial and regulatory practices, and because Chinook numbers are declining. In 2007, the Chinook bycatch in the pollock-directed fisheries was 121,770 fish, with an additional 7,798 Chinook bycatch in other Bering Sea groundfish fisheries. In 2012, the numbers had been dramatically reduced, with 11,352 Chinook incidentally caught in the pollock-directed fisheries.[32]

The 2010 Bering Sea/Aleutian Island pollock catch was worth in excess of $282 million, and products from pollock were worth in excess of $1 billion. By comparison, the Chinook bycatch had a potential economic value that cannot be accurately estimated, in part because a portion of the Chinook harvest is used outside of the Western economy and is subject to different cultural logics. Mangers often impose a Western economic analysis on the value of subsistence-caught fish—the neoliberal solution—but to do so elides local cultural significance and obscures the interplay of market-based and nonmarket-based economies. Estimates of the "replacement value" of subsistence foods, including Chinook salmon, have been advanced, but the values range from $3.50/pound to $118/pound, based on different methods of valuation, all of which are couched in terms of Western economic models. The large differences in per pound valuations indicate, at a minimum, the lack of scholarly consensus on such estimates. More likely, the large differences indicate the inapplicability of Western economic models to non-Western economies. Despite the potential inapplicability of such models, they continue to be developed and used in natural resource policymaking.[33]

As administered by the CDQ program, benefits to western Alaska villages flowing from the pollock fishery have been mixed. By 2007, royalties paid to CDQ program participants were $50.3 million. By 2008, total revenues for CDQ participants exceeded $190 million. By 2013, the combined net worth of the six nonprofit corporations established to administer the program exceeded $785 million. Residents of the 65 participating native villages, many of whom depend upon Chinook salmon for subsistence, profit from the economic development made possible by the CDQ program in the form of subsidies for boat motors, college scholarships, and heating fuel assistance. Yet the range of senior executive salaries for these nonprofits—from $69,503 to $832,367—stands in sharp contrast to the average commercial income from Yukon River fishers, which in 2009 was under $2,000. In addition, the catch-share program has not evenly distributed its resources to the six nonprofit corporations, and thus to local villages, resulting in conflict between the corporations.[34]

The central problem for subsistence fishers, however, is that the pollock fishery is in part responsible for Chinook declines. In fact, part of the Bering Sea/Aleutian Islands Chinook salmon bycatch is allocated to the CDQ program. In other words, CDQ participation in the Bering Sea pollock fishery may have had the contradictory effect of undermining traditional subsistence practices in the pursuit of economic development goals.[35]

Any potential conflict has been rarely discussed. But such conflict is becoming more publicly prominent as CDQ nonprofit organizations lobby Congress to alter catch-share apportionment, and as local subsistence fishers conduct protest fisheries in the face of emergency Chinook salmon fishery closures. Debates about subsistence practices vs. development goals have ensued.[36]

The 2009 Bering Sea Chinook Salmon Bycatch Environmental Impact Statement summarizes, and perhaps understates, the problem of environments, management, and markets:

> Management of the Yukon salmon fishery is difficult and complex because of the often inability to determine stock specific abundance and timing, overlapping multi-species salmon runs, increasing efficiency of the fishing fleet, the gauntlet nature of the Yukon fisheries, allocation issues between lower river and upper river Alaskan fishermen, allocation and conservation issues between Alaska and Canada, and the immense size of the drainage.[37]

Customary Trade: Non-Western Economies

In broad outline, these are the intertwined contexts within which to situate small-scale exchanges of cash for subsistence-caught Chinook salmon along the Yukon River: scales of harvest, laws and regulations, managed environments, markets, and declining populations of Chinook. State and federal programs have uneasily accommodated small-scale cash exchanges in part because, to borrow James Scott's term, they have not been "legible" to bureaucrats. No one knows the magnitude of customary trade. As a consequence, it has remained mostly unregulated and the subject of increasing controversy.

The available ethnographic record, however, provides glimpses into the practice. Anthropologist Ann Fienup-Riordan describes cash sales of locally harvested foods occurring in the early 1980s in the Yup'ik-speaking communities of Alakanuk, Sheldon's Point, and Scammon Bay near the mouth of the Yukon River. These and other communities in the area had developed a system of interregional sale and trade of subsistence foods, including salmon. As Fienup-Riordan recognizes, to analyze exchanges of cash for subsistence foods requires the theoretical armamentarium of anthropology and not of economics. Kinship, exchange practices, local meanings, relations of humans to animals, cosmology—these elements of cultural life, not easily accessible to economic thought, offer insight into how cash exchanges are culturally motivated. The following descriptions are from Fienup-Riordan.[38]

The extended family, consisting of several households connected by marriage and descent relations, is the primary unit of production within Yup'ik-speaking communities of the lower Yukon delta. Households that make up the extended family may occupy a single village or be spread across several villages. Household harvest efforts, and the distributions of harvests, are organized by cultural norms. Households may specialize in particular harvests—fishing for salmon, for example, or hunting for seals—but the goal is not household independence; the goal is interdependence of the extended family network. Interdependence is expressed in a variety of ways. These include gender-defined tasks, age-related distributions of harvests, development and maintenance of social hierarchies, and the strong desire to ensure that everyone within the extended family is sufficiently provisioned with diverse, locally harvested foods.

Subsistence foods always have higher value than store-bought commodities. Minimizing effort and maximizing efficiency in the pursuit of these foods are secondary considerations to the value placed on them and may not be considered at all given the need to adequately provide for everyone within the extended family. Extra effort and expense are often needed to provide locally harvested foods for distribution throughout the network. One means to compensate a family member for the extra effort and expense is through an exchange of cash. Such exchanges are not for food but for the effort expended to procure it and are intended to ensure the effort itself will continue.[39]

Distributions of harvests are controlled by older women. "Food stores," writes Fienup-Riordan, "especially staples such as salmon and seal oil, are often kept in a common food cache behind the parent's house." The result of efforts of many people within the extended family, the common food cache nonetheless cannot be accessed by those people.

> The oldest woman of the extended family group keeps a careful eye on this common larder, and has a shrewd knowledge of what has been contributed and by whom. She is the one to decide what is to be eaten and when. Although this cache is the product of the joint effort of the extended family unit, draws of dried fish, oil, and herring by younger householders take on the character of a request, and in fact do not occur at all if the stores are limited. Once the stores have been accumulated, the majority through intense efforts of the men, they become the property of the older women of the extended family household for processing and for distribution within and ... beyond that unit.

Such distributions of harvests also generate unequal relations of power. Culturally defined status results from generosity, from the ability and practice of giving to those with fewer resources; it does not result from the accumulation of wealth. Among Yup'ik, the practice of sharing is widespread and includes relatives, nonrelatives, other villagers, friends, and strangers. "All manner of goods are exchanged," notes Fienup-Riordan, "both the scarce and the plentiful, the valuable and the ubiquitous." The system can be summarized in the local saying, "You are really rich if you eat only gifts, and give all you have away."

Status is double-edged, however. Fienup-Riordan cites the local aphorism, "Gifts make slaves as whips make dogs," and explains it as follows:

> [T]he gift becomes the mechanism for the establishment of a power hierarchy. This aspect of the ubiquitous shared meal and gift of shared meat should never be underestimated. The contemporary village can, in fact, be understood as a collection of overlapping extended family networks, wherein the most elaborate gift giving is accomplished by the most wealthy, and

correspondingly powerful, networks. These same extended family groups are those which invest the largest percentage of their incomes into harvesting pursuits. Yet they are sometimes difficult to distinguish, as in worldly goods they do not tower above their peers. This is the result of the support they supply to less well provisioned family networks. Although difficult to measure, redistribution of the harvest is a critical element in the economy as well as the social hierarchy of [Yukon River] delta communities. At the same time that it valorizes social distance, it diminishes economic discrepancies, with wide ranging implications for the significance of the harvesting process.

One such implication is the cultural significance of cash exchanges. In 1982, a spotted seal sold for $40 in Scammon Bay and was resold in Mountain Village for $60. Sheefish sold for between three and seven dollars, depending upon size. Six gallons of dried chum salmon sold for between $100 and $150. The purpose of such sales, however, was not to make a profit or to accumulate wealth. The purpose was to circulate food through networks of kin.

> In all of these cases, although the transaction was consummated with cash, the primary motive in the harvest of the resource was not strict economic gain. Only a handful of households in each village produce extra salmon or harvest extra seals specifically for sale. The majority of households sold or traded irregularly, only in the case of an unusually large harvest … In fact, in the event of an abundant supply, what happens in the majority of cases is not the conversion of the excess to economic value, but the extension of the effective kin group through the distribution of the catch.

Fienup-Riordan emphasizes that it is a mistake to interpret sales of subsistence-caught foods as commercial in nature and to impose a set of Western economic values on transactions that have other cultural logics.

> The social justification for what might otherwise be interpreted as an activity undertaken for profit brings us back to the original goal of the exchange system, that is: to accumulate within the extended family for distribution beyond it, both within the village and between villages, at whatever level the individual household or extended family can maintain.

The larger point is that sharing of subsistence foods, as anthropologist Tim Ingold argues, "constitutes the common purpose that people *bring into* the productive process itself."[40] Sharing, in other words, is not the end of production; it is the very reason for production. People engage in subsistence harvests and distributions to meet their social obligations to others. The majority of those obligations are based on kinship. Cash exchanges are embedded in such relations, which remain more significant than cost/benefit calculations, supply and demand, or other market forces, and certainly more difficult to quantify. These exchanges are not "commercial" in nature but are seen from the outside as such only because cash is involved in the transaction (this point will come up again in terms of the health safety of traditionally processed foods).

Nor do obligations end with one's kin, neighbors, and friends. Such obligations extend to the animals one kills. Among Yup'ik, all creatures are considered to be sentient nonhuman persons endowed with kinship ties, language, desires, and the ability to control their own destinies. This is a widespread understanding for many northern peoples and results in relationships to animals often difficult for others to comprehend. Animals are thought to willingly give themselves to hunters; hunters in turn do not own harvested animals but are obligated to pass on the catch. The involvement of cash does not alter cultural attributions of the exchange of harvested animals, at least not to the extent of transforming them into significant commercial enterprises.[41]

Other ethnographic studies come to similar conclusions. In 2004 and 2005, Catherine Moncrieff interviewed 28 active fishers from three Yukon River communities: Alakanuk, Holy Cross, and Tanana. In Alakanuk, fishers reported that a few villagers sold subsistence-caught salmon, but in limited quantities, which ranged from quart-sized bags of smoked salmon strips for $20 apiece to five-gallon buckets of dried chum salmon for $200 apiece. One interviewee noted he had sold subsistence-caught salmon for 20 years, but only if he had extra fish to sell beyond the needs of his family. Another interviewee noted that he sold or traded salmon with people in several communities, among them Hooper Bay, Chevak, Scammon Bay, Stebbins, and Anchorage. One Alakanuk resident reported buying subsistence-caught salmon: a box of dried chum for $40. Moncrieff's study indicates that the reasons residents of Alakanuk sold subsistence foods were similar to those Fienup-Riordan reported some 20 years earlier: to help those who could not fish, to avoid waste, and to generate cash to enable further subsistence activities. As in the early 1980s, so in the early 2000s: customary trade comprised an important but relatively modest aspect of the local subsistence economy.[42]

Further up the Yukon River, at the village of Holy Cross, residents more frequently sold subsistence-caught fish, which they processed in a variety of ways. Quantities varied year by year. One interviewee sold 18 salmon processed into six cases of pint jars. Other interviewees sold an average of 30 to 40 pounds of salmon, with prices depending upon species and quantity. Chinook salmon strips sold for $20 per quart bag or $16–$20 per pound. Half-dried salmon bellies sold for $75 per case. Moncrieff found it difficult to obtain total sales numbers, but on average interviewees appear to have sold $1,360 worth of salmon in customary trade per year. Cash generated from these sales was used to purchase gas and supplies for subsistence activities, household items, children's clothing, and to pay utility bills. Moncrieff concludes that cash obtained from the customary trade of salmon made further subsistence fishing possible, and provided small amounts of money for other household expenses.[43]

In Tanana in 2005, Moncrieff's interviewees indicated that they sold subsistence-caught salmon in amounts ranging from a few fish to 100 Chinook salmon. One person had sold an average of 600 pounds of salmon for $6,000 annually, but in 2005 had reserved most of his harvest to share with a large network of family and friends. Fish were sold to family and friends in other communities, including Manley Hot Springs, Nenana, Fairbanks, Salcha, Sitka, Minto, Minchumina, Ruby, Point Hope, and elsewhere. Most of the salmon were sold as dried strips or as dried fish but were available in a variety of processed forms. Prices were fairly consistent:

> Whole fish, $1/pound
> Fillets, $2/pound
> Half-dried fish, $5/pound
> Strips of dried fish, $15–$18/pound
> Dried fish, $12–$18/pound
> Canned strips of fish, $12–$15/tall can
> Canned fresh fish, $6/short can, $15/tall can, $8/jar

Moncrieff did not report the salmon species associated with these sales, nor the amounts earned from them, but noted that study participants used the income from customary trade to fund subsistence fishing activities.[44]

Fishers interviewed in Moncreiff's study reported that they engaged in customary trade only if they first harvested sufficient fish for their own family's needs and satisfied their obligations to share fish with a network of extended family and friends. They did not subsistence fish primarily to sell fresh or processed salmon. Cash raised through customary trade appears to support other subsistence activities and is used to pay for household and other expenses. Generating profits or operating a business were not characteristics of customary trade as reflected in interviews with residents of Alakanuk, Holy Cross, and Tanana.

Two other studies of customary trade bear mention, although they are not focused on the Yukon River. The first is a study in the Bristol Bay area south of the Yukon River, the second is a study in the Seward Peninsula area north of the Yukon River.

Theodore Krieg and colleagues describe sharing, barter, and cash trade among Central Yup'ik speakers in the Bristol Bay area in 2004. Their study included the communities of Dillingham, Togiak, Naknek, King Salmon, and Nondalton. Perhaps the most useful aspect of this study was the development of a Central Yup'ik typology of sharing and trade, which indicates the difficulty translating the regulatory terms "sharing," "barter," and "customary trade" into Central Yup'ik. The nuances between forms of sharing also indicate the cultural importance of the practice. A variety of Central Yup'ik terms distinguish between "giving material things," "receiving gifts without hesitation," "distributing a share of the catch," "taking food to someone," "sharing an item out of respect for an invitation," "setting aside food for someone," "trading," "borrowing," and "to sell for money," among others.[45]

Tune-, "to sell for money," is a rough translation of "customary trade." As the researchers point out, "[i]n this form of exchange, the monetary price of an item is set largely by custom...." Moreover, "[t]here is no haggling over price, no profit motive, and no competition for sales between producers." *Tune-* has also been extended to refer to market sales, but its primary meaning appears to refer to exchanges for cash that have been set by custom and not by market forces. As in Fienup-Riordan's study of the Yukon Delta area, exchanges for cash in the Bristol Bay area were motivated "by an interest in maintaining relationships, providing a service to friends and family, and preserving community traditions." Krieg and colleagues concluded that "the amount of salmon involved in barter or cash trade… is likely quite small in comparison with other fishing activities, including the total subsistence harvest." The researchers also provided a caution: "Attempts at regulating these activities may be counterproductive."[46]

In 2004–2006, James Magdanz and colleagues studied customary trade and barter in the Seward Peninsula area and developed a network analysis that illustrates the extent to which subsistence foods are distributed throughout the area. Communities in this study included Brevig Mission, Elim, Nome, Saint Michael, Shaktoolik, and Stebbins.[47]

As in the Yukon Delta, the lower and middle areas of the Yukon River, and Bristol Bay, people in the Seward Peninsula area engaged in cash exchanges for subsistence foods as a normal part of their lives. Such exchanges were infrequent and involve small sums of money. Of the $7,806 recorded in cash trades, 44 percent were for less than $50. Ninety percent of the 88 households surveyed reported annual trades of less than $500. One finding in the Seward Peninsula study stands out as distinct from other areas. Trade relationships involving cash were short term: "customary trade relationships were more likely to be economically motivated, not personally motivated…people seemed more likely to sell fish to people they had met recently." Nonetheless, "Whether selling to friends or strangers, profit did not seem to be the motive."[48]

Most customary trades remained within Northwest Alaska. Interregional exchanges were rare. Six percent ($450) of cash exchanged for local subsistence foods came from outside Northwest Alaska. One percent (30 pounds) of subsistence foods exchanged for cash came from outside of Northwest Alaska.

Magdanz and coauthors conclude their study by pointing out the arbitrary division between small-scale commercial and noncommercial fisheries in Alaska. Unlike Canada and Greenland, which have incorporated subsistence foods—"country foods"—into local cash economies, the State of Alaska has elected to regard "all cash exchanges of fish, regardless of scope, as commerce, and has prohibited most cash exchanges of wildlife, except for trapping." The result has been the development of small-scale freshwater commercial fisheries, "commercial in name only." By way of example, Magdanz and his colleagues cite a fishery for sheefish "conducted with gillnets set under the ice in winter, a traditional subsistence method." This commercial fishery developed out of a subsistence fishery in the late 1960s, but in the last ten years, only one fisher participated and sold about 100 fish per year.

It could be argued that some of the small "commercial" fisheries in Northwest Alaska met the definition of "customary trade," except the Department [of Fish and Game] elected to manage them as commercial fisheries. The advantages (for the Department) were that the freshwater "commercial" fisheries were easy to establish administratively, and did not require Board of Fisheries approval. Licensing, reporting, and other systems were all in place for "commercial" fisheries. In practice, though, subsistence fishers often found that the $75 commercial license and the fish tickets were not worth the hassle for a hundred fish. Eventually, as the Kotzebue sheefish fishery may illustrate, many opted out of the commercial management system, but continued to fish for subsistence, and perhaps to quietly sell a few fish to their neighbors.[49]

Despite such studies, neoliberal policies remain blind to the cultural significance of natural resources, simply because such significance cannot be reduced to pecuniary considerations. As neoliberal policies become more firmly established, local cultural practices that contradict those policies are ignored or, at the other extreme, criminalized.

Health Safety of Traditionally Processed Salmon

Selling a few fish to neighbors, to family members who live in distant communities, or to newcomers to one's community who cannot otherwise procure fish, may characterize customary trade. But the federal subsistence program, as it struggled with understanding the practice in 2001 and over the next decade, had to contend with other issues, including the state's assertion that processed seafoods exchanged for cash must meet Department of Environmental Conservation health safety standards. The assertion is not surprising. The state has a stake in ensuring the safety of foods that are marketed. Most of the Chinook salmon that are exchanged in customary trade have been traditionally processed, typically as smoked and dried "strips." Are these traditionally processed foods subject to state health safety oversight? If so, the state has a legal mechanism to close down customary trades, even as they are allowed by federal statute. The state, however, has not often engaged this mechanism, apparently preferring to ignore activities it asserts are illegal. The lack of enforcement provides the occasion to question whether health safety is at issue or whether some other tacit concern motivates the state's position.

An October 28, 2002 letter to the Federal Subsistence Board from Janice Adair, Director of Environment Health, Department of Environmental Conservation, State of Alaska, appears to be the authority upon which assertions of state oversight rely. In that letter, Adair states that "Alaska has the highest rate of foodborne botulism in the United States and *all* cases have been associated with traditional Alaska Native foods." She urges the Federal Subsistence Board to enact regulations that allow only fresh or frozen fish to be exchanged under customary trade. Adair also acknowledges that "I am not at all certain regarding the interplay between the proposed regulations [on customary trade] and state law."[50]

What she fails to point out, however, is that traditionally processed, smoked and dried salmon has been implicated in only a few botulism cases; most cases are associated with traditionally "fermented" fish eggs and fish heads, and fermented parts of marine mammals (the process is often referred to as "fermentation," but it is more accurately described as "putrefaction"). Between 1947 and 2007, out of 131 cases of botulism outbreaks, two involved dried fish (species unidentified, but one case was of spoiled fish). Whether traditionally processed Chinook salmon "strips" were involved is unknown.[51]

Nor does the state assert that "bartered" or "shared" strips are subject to the same health standards. A subsistence fisher can barter or share traditionally processed Chinook salmon under federal rules, but the exchange of cash *for the very same fish* causes the state to take notice. With the involvement of cash, the state's interest shifts from the empirical to the ideological as it imposes a set of market values and restrictions onto nonmarket trades, but does so in the guise of protecting the well-being of its citizens. Cash appears to have the magical value of catching the state's attention. Barter and sharing do not.

Yet, in a contradiction characteristic of bureaucratic effort, the state has not asserted that health regulations govern the sale of marine mammals, even though the sale of subsistence-caught marine mammals is allowed under the Marine Mammals Protection Act (an act the Federal Subsistence Board does not administer). This act states that "any edible portion of marine mammals may be sold in native villages and towns in Alaska or for native consumption," provided the taking was for subsistence purposes by "any Indian, Aleut, or Eskimo who resides in Alaska and who dwells on the coast of the North Pacific Ocean or the Arctic Ocean" (16 USC 1371).[52] The state has not claimed that sales of subsistence-caught marine mammals such as bowhead whale or walrus must meet Department of Environmental Conservation health safety standards. Indeed, it is difficult to imagine a health inspector on the beach in Barrow, Alaska ensuring that the flensing of a bowhead whale is conducted according to bureaucratic guidelines intended to maintain standards of hygiene. Fish, however, appear to be different; that difference involves markets.

The exchange of subsistence-caught food for cash is easily misinterpreted as a market phenomenon. The state statute that governs the processing and sale of fish refers to "products to be sold as part of commerce" (19 AAC 34). The bureaucratic question is whether customary trade, in which whole or processed fish are exchanged for cash, constitutes either a "sale" or is an element of "commerce" within a market economy.

The ethnographic record indicates that customary trade and related practices described as sharing and barter stand outside of the market economy. In his draft report entitled "Customary Trade in Alaska: An Introduction," anthropologist Robert Wolfe notes that "Customary trade is one way that wild foods are distributed through non-commercial channels between households in rural Alaska…." He adds that "Studies by the [Alaska] Division of Subsistence indicate that subsistence foods commonly flow from producers to consumers through non-commercial distribution systems."[53] Wolfe cites a number of studies to support his statement, to which other studies lend further empirical evidence.[54]

The state statutes contain several exemptions to health safety standards (18 AAC 31.012). For example, seasonal events such as fairs and bazaars, at which bakery items are sold, are exempted from state health regulations. Homemade foods and other items sold to a consumer by the individual who prepared the items are exempted. Farmers' markets and roadside stands are also exempted if the items sold are raw fruits and vegetables in their natural state or with minimal processing. Cultural organizations—clubs, churches, Alaskan Native potlatches—as well as informal gatherings such as parties, picnics, and potlucks held by an office, a family, neighborhood, or community, are similarly exempted from state health regulations.

Most of these and comparable exemptions to state health regulations focus on nonmarket activities that have long been a part of the broader American culture. Strong associations with an agrarian past (roadside stands, fairs, farmers' markets) and with formal and informal cultural activities practiced by churches, other religious organizations, neighborhoods, and families, figure prominently in state exemptions.

In developing its regulations on customary trade, the federal program did not suggest that salmon processed in traditional ways and exchanged for cash was analogous to the perhaps more familiar cultural practices exempted under state health regulations. Instead, the federal program developed regulations governing customary trade and then promptly, albeit informally, negated them.

The federal regulations distinguish between rural and nonrural residents (Code of Federal Regulations, CFR 50 § 100.27. Subsistence taking of fish). "Rural residents may exchange in customary trade subsistence-harvested fish, their parts, or their eggs, legally taken under the regulations in this part, for cash from other rural residents." For transactions between rural and nonrural residents, the rules are different.

> In customary trade, a rural resident may trade fish, their parts, or their eggs, legally taken under the regulations in this part, for cash from individuals other than rural residents if the individual who purchases the fish, their parts, or their eggs uses them for personal or family consumption.

In both situations, fish, their parts, and their eggs can be exchanged for cash. To render fish into parts and eggs requires some kind of processing.

Yet in the pamphlet, *Subsistence Management Regulations for the Harvest of Fish and Shellfish on Federal Public Lands and Waters of Alaska*, the federal subsistence program handbook most rural residents refer to, the message was reversed and the point emphasized in bold lettering: "In practical terms, the only type of customary trade allowable for those who do not process their fish in accordance with state food safety regulations is the sale of uncut, unprocessed fish."[55]

Thus state regulations, concerned with the health safety of processed fish, have been asserted to apply to sales of subsistence-caught fish conducted under federal regulations, rendering such sales illegal with the exception of whole, unprocessed fish—in effect, contradicting the federal rule allowing the sale of fish, parts, and eggs, and turning long-standing customs into illegal acts. But that assertion may be a red herring, regardless of how many times, or which authority, repeats it. Although repeated for over a decade, claims that state food safety regulations superseded federal regulations have never been tested in court. Moreover, state regulations apply to processed foods that enter the stream of commerce. They do not apply to foods sold outside of the market economy and that do not enter the stream of commerce. In any event, the state regulations are unevenly applied. They focus on fish and not on other processed subsistence foods—bowhead whale, walrus, seal, moose, caribou—that are exchanged for cash. They also fail to recognize the congruence between exempted practices under state law and customary trade. To borrow the U.S. Ninth Circuit Court of Appeal's phrase, cited above, the "transparent purpose" of the state's assertion "is to protect commercial…interests"—interests that would suffer in the event botulism was associated with any wild Alaska salmon.

And suffer the salmon industry did in the early 1980s, after the death of a Belgian man who was poisoned with botulism from eating salmon from a Ketchikan, Alaska cannery. The death prompted the second-largest recall of canned goods in U.S. history and threatened to undermine the industry. The state, in arguing that its health safety regulations apply to customary trade, is taking preemptive action against future threats to the seafood industry, even if the regulations do not apply. Recognizing and allowing long-standing cultural practices are of secondary concern.[56]

For its part, the federal program appears to have both agreed and disagreed with the state: agreed, by adopting in its publications and at public meetings the state's position on health safety, and disagreed, by allowing customary trades of fish parts and eggs, state health warnings notwithstanding. Rural residents are thus encumbered by rules which they nevertheless proceed to break, if not with enthusiasm, then with the aim of preserving their own relations with the natural world and each other.

"You Can't Just Let Nature Run Wild"[57]

A well-ordered fishery is one in which all fish and all harvests are accounted for. Such accounting in turn allows managers to allocate catch across different user groups—subsistence, sport, and commercial—and to control those groups by specifying gear types, fishing time, and limits. Control is further characterized by regulations, permits, enforcement practices, management plans, record-keeping, and the development of datasets—all of which are used to inform managers and improve the efficiency of control. A well-ordered fishery attempts to eliminate uncertainty and only poorly accommodates practices that do not adhere to its rationality.

The Yukon River Chinook salmon fishery is not well-ordered. Neither natural nor social systems have proven amenable to simple rationalities of control. The numbers of returning Chinook are not known with any certainty, and escapement goals, set by biologists, are open to question. Local cultural practices, which persist despite the intrusion and pervasiveness of market forces, have proven difficult for managers to understand. Under these conditions, harvests cannot be equitably distributed. The effects of a changing climate, contests between state and federal systems, a diversity of peoples speaking different languages and engaged in fundamentally different kinds of economies, and the vagaries of the open ocean fishery, in which neoliberal policies have become increasingly dominant—all of these combine to thwart any efforts of control.

Written more than 40 years ago, ANILCA contains language intended to preserve elements of local culture by promoting economic activities such as customary trade that support cultural continuity. Congress perceived these activities as related to "barter" and "sharing," both of which stand outside of markets. Through ANILCA, Congress refused to reduce cultural forms to market values, even though, through ANCSA, it had done just that. Recently, however, customary trade has become increasingly controversial and subject to greater regulatory pressure because of its presumed relation to markets. The exchange of cash for subsistence-caught fish has become the concern of four interrelated groups of people: those who have a stake in markets, such as commercial fishermen; those who regulate markets, such as state entities charged with promulgating health regulations governing sales of processed foods; those who manage fisheries, whose management is made more difficult by exchanges that are opaque to managerial scrutiny; and those who are charged with enforcing regulations that many people do not understand and, in any case, largely ignore.

As a general historical phenomenon, involvement in the market economy results in ever-increasing involvement, leading to a dependency on that economy and a loss of subsistence capacity.[58] In Alaska, however, subsistence practices have been maintained in part because of the value placed on subsistence by local cultures and in part because government regulations have allowed some subsistence activities to continue on federal lands and waters. The great surprise of the early twenty-first century, as anthropologist Marshall Sahlins has noted of late capitalism generally, is that, despite the encroachment of the capitalist economy, many hunters and gatherers persist—and they do so by hunting and gathering and, in our context, by fishing.[59]

Or, increasingly, they persist not only by fishing but by trading cash for fish. Jack Reakoff, a prominent subsistence advocate in Alaska, put the matter as clearly as it can be put. Trading cash for fish is "how fish is disseminated throughout the region away from the river." He goes on to comment, "whether the federal government can tolerate it or the state can tolerate it, we consider that as customary use…it's just the way it works."[60]

Notes

1 For these and related stories, see Eastern Interior Alaska Subsistence Regional Advisory Council transcripts, Volume 11, February 25, 2010, 269 ff. U.S. Fish and Wildlife Service, Office of Subsistence Management. Anchorage, AK. Yukon-Kuskokwim Delta Alaska Subsistence Regional Advisory Council transcripts, February 23, 2012, 107. U.S. Fish and Wildlife Service, Office of Subsistence Management. Anchorage, AK.

2 These numbers combine "personal use" and "subsistence" categories. James A. Fall, et al., *Alaska Subsistence Salmon Fisheries 2009 Annual Report* (Alaska Department of Fish and Game Division of Subsistence, Technical Paper No. 373 Anchorage, 2012), 7, 39. Douglas M. Eggers, Michael D. Plotnick, and Amy M. Carroll, eds., *Run Forecasts and Harvest Projections for 2010 Alaska Salmon Fisheries and Review of the 2009 Season* (Anchorage, AK: Alaska Department of Fish and Game, Special Publication No. 10-02, 2010), 22.

3 Eggers, Plotnick, and Carroll, eds., *Run Forecasts and Harvest Projections for 2010 Alaska Salmon Fisheries and Review of the 2009 Season*, 22.

4 *Subsistence in Alaska: A Year 2010 Update* (Anchorage, AK: Alaska Department of Fish and Game, Division of Subsistence, 2012). *Alaska Population Overview: 2012 Estimates* (Alaska Department of Labor and Workforce Development, Research and Analysis Section, November 2013).

5 On neoliberalism, see David Harvey, *A Brief History of Neoliberalism* (Oxford: Oxford University Press, 2005); Bob Jessop, *The Future of the Capitalist State* (Malden, MA: Polity Press, 2002); Noam Chomsky, *Profit Over People: Neoliberalism and Global Order* (New York, NY: Seven Stories Press, 1999); Jean Comaroff and John L. Comaroff, eds., *Millennial Capitalism and the Culture of Neoliberalism* (Durham, NC: Duke University Press, 2001). On the inapplicability of western economic models for non-market economies, see John M. Gowdy, ed., *Limited Wants, Unlimited Means: A Reader on Hunter-Gatherer Economics and the Environment* (Washington, DC: Island Press, 1998). James C. Scott, *Seeing Like a State: How Certain Schemes to Improve the Human Condition Have Failed* (New Haven, CT: Yale University Press, 1998), is central to this discussion, and I've adopted some of his points in this paragraph.

6 Title 5 Chapter 01.717 of the Alaska Administrative Code allows the sale of subsistence-harvested herring roe on kelp in Southeast Alaska, and Title 5 Chapter 01.188 allows the sale of subsistence-harvested finfish in the Norton Sound-Port Clarence area. George E. Pappas, *1974–2010 Customary Trade of Subsistence Caught Fish: Background, Chronology, and Current Options for Modification* (Anchorage, AK: Alaska Department of Fish and Game, Office of the State-Federal Subsistence Liaison Team, 2012).

7 David S. Case and David A. Voluck, *Alaska Natives and American Laws*, second ed. (Fairbanks, AK: University of Alaska Press, 2002).

8 Customary trade was subsequently linked to "personal and family needs." 64 *Federal Register* 1287; January 8, 1999.

9 Senate Report No. 413, 96th Congress, 2nd Session, 234.

10 On cash and culture, see generally Maurice Block and Jonathan Parry, *Money and the Morality of Exchange* (Cambridge: Cambridge University Press, 1989). Useful specific studies include Paul K. Eiss, "Hunting for the Virgin: Meat, Money and Memory in Tetiz, Yucatán," *Cultural Anthropology* 17:291–330 (2000); Michael Nihill, "The New Pearl Shells: Aspects of Money and Meaning in Anganen Exchange," *Canberra Anthropology* 12:140–160 (1989); Mahir Şaul, "Money in Colonial Transition: Cowries and Francs in West Africa," *American Anthropologist* 106:71–84 (2004); Jeffery G. Snodgrass, "A Tale of Goddesses, Money, and Other Terribly Wonderful Things: Spirit Possession, Commodity Fetishism, and the Narrative of Capitalism in Rajasthan, India," *American Ethnologist* 29:602–636 (2002); and Michael T. Taussig, *The Devil and Commodity Fetishism in South America* (Chapel Hill, NC: The University of North Carolina Press, 1980). Niall Ferguson's popular *The Ascent of Money: A Financial History of the World* (New York, NY: Penguin Press, 2008), offers an account untroubled by anthropology. Compare David Graber, *Debt: The First 5,000 Years* (London: Melville House, 2011), which is resolutely anthropological.

11 Alaska Stat. Sec. 16.05.940(25) (1987).

12 Kenaitze Tribe v. State of Alaska, 860 F.2d 312 (1988).

13 McDowell vs. State of Alaska, 785 P.2d 1 (Alaska, 1989).

14 55 *Federal Register* 26114 (June 29, 1990).

15 Harvey, *A Brief History of Neoliberalism*. See also S. Ganapathy, "Alaskan Neo-Liberalism: Conservation, Development, and Native Land Rights," *Social Analysis* 55:113–133 (2011).

16 See J.E. Morrow, *The Freshwater Fishes of Alaska* (Anchorage, AK: Alaska Northwest Publishing Company, 1980), and Charles C. Kreuger and Christian E. Simmerman, eds., *Pacific Salmon: Ecology and Management of Western Alaska's Populations* (Bethesda, MD: American Fisheries Society, 2009). Dan O'Neill provides an engaging and accessible account of how and why Chinook (King salmon) runs declined on the Yukon River, "The Fall of the Yukon Kings," in *Arctic Voices: Resistance at the Tipping Point*, S. Banerjee, ed. (New York, NY: Seven Stories Press, 2012).

17 David B. Anderson and Cheryl L. Scott, "An Update on the Use of Subsistence-Caught Fish to Feed Sled Dogs in the Yukon River Drainage, Alaska," in *Final Report 08-250* (Anchorage, AK: U.S. Fish and Wildlife Service, Office of Subsistence Management, Fisheries Resource Monitoring Program, 2010).

18 O'Neill, *The Fall of the Yukon Kings*, 148–149.

19 Alaska Department of Fish and Game Team. "Chinook salmon stock assessment and research plan, 2013" (Anchorage, AK: Alaska Department of Fish and Game Special Publication No. 13-01, 2013), 34 ff. "Treaty between the Government of Canada and the Government of the United States of America Concerning Pacific Salmon." First signed in 1985, this treaty has been periodically updated. www.psc.org/about-us/history-purp ose/pacific-salmon-treaty/.

20 Federal Subsistence Board meeting transcripts, December 11, 2007. Office of Subsistence Management (Anchorage, AK), 89. Federal Subsistence Board Meeting Materials, January 2012. Office of Subsistence Management (Anchorage, AK), 163.

21 Alaska Department of Fish and Game recognizes "biological escapement" goals intended to "achieve maximum sustainable yield (human use)," which are set by ADF&G; "optimum escapement goals," which "allow for sustainable runs based on biological needs of the stock and ensures healthy returns for commercial, sport, subsistence, cost-recovery, and personal use harvests," which are set by the Board of Fisheries; and "sustainable escapement goals," which are estimates "based on historical performance and other factors known to conserve stock over a five to 10 year period," and which are developed by ADF&G "taking into account data uncertainty." See Amy Carroll, "What Are Escapement Goals?" *Alaska Fish and Wildlife News* (February 2005).

22 See Robert Clark, M. Willette, S. Fleischman, and D. Eggers, "Biological and Fishery-Related Aspects of Overescapement in Alaskan Sockeye Salmon *Oncorhynchus nerka*," Alaska Department of Fish and Game, Special Publication No. 07-11 (Anchorage, AK, 2007); compare Thomas B. Dunklin, "Over-Escapement: Is There a Problem?" Wild Salmon Center (February 2007). www.wildsalmoncenter.org/pdf/Overescapem ent_Thomas_Dunklin_2005.pdf. On nutrient loads, see John Stockner, ed., *Nutrients in Salmonid Ecosystems: Sustaining Production and Biodiversity*. American Fisheries Society, Symposium 34 (Bethesda, MD: American Fisheries Society, 2003).

23 Case and Voluck, *Alaska Natives and American Laws*, 291.

24 The quote is from. Douglas M. Eggers, Michael D. Plotnick, and Amy M. Carroll, eds., *Run Forecasts and Harvest Projections for 2013 Alaska Salmon Fisheries and Review of the 2012 Season* (Anchorage, AK: Alaska Department of Fish and Game, Special Publication No. 13-03, 2013). See also Philip A. Loring and Craig Gerlach, "Food Security and

Conservation of Yukon River Salmon: Are We Asking Too Much of the Yukon River?" *Sustainability* 2:2965–2987 (2010).

25 Alaska Department of Fish and Game Team. Chinook salmon stock assessment and research plan, 2013. Alaska Department of Fish and Game Special Publication No. 13-01 (Anchorage, AK, 2013).

26 Dean Bavington, *Managed Annihilation: An Unnatural History of the Newfoundland Cod Collapse* (Vancouver, BC: University of British Columbia Press, 2011).

27 For an overview, see Eggers, Tide, and Caroll, eds. *Run Forecasts and Harvest Projections for 2013 Alaska Salmon Fisheries and Review of the 2012 Season.*

28 Becky Mansfield, "Neoliberalism in the Oceans: 'Rationalization,' Property Rights, and the Commons Question," *Geoforum* 35:313–326 (2004). Becky Mansfield, "Rules of Privatization: Contradictions in the Neoliberal Regulation of North Pacific Fisheries," *Annals of the Association of American Geographers* 94:565–584 (2004). Fishery Management Plan for Groundfish of the Bering Sea and Aleutian Islands Management Area (Anchorage, AK: North Pacific Fishery Management Council, June 2012), 43. Casey Grove, "Two Bering Sea Pollock Processors Lied about Catches, Feds Say," *Anchorage Daily News* (May 17, 2013).

29 77 Federal Register 42629; July 20, 2012.

30 See Jeremy B.C. Jackson, M.X. Kirby, W.H. Berger, K.A. Bjorndal, L.W. Botsford, B.J. Bourque, R.H. Bradbury, R. Cooke, J. Erlandson, J.A. Estes, T.P. Huges, S. Kidwell, C.B. Lange, H.S. Lenihan, J.M. Pandolfi, C.H. Peterson, R.S. Steneck, M.J. Tegner, R.R. Warner, "Historical Overfishing and the Recent Collapse of Coastal Ecosystems," *Science* 293:629–637 (July 27, 2001); David O. Conover and Stephan B. Munch, "Sustaining Fisheries Yields over Evolutionary Time Scales," *Science* 297:94–96 (July 5, 2002), and Christian Jørgensen, K. Enberg, E. Dunlop, R. Arlinghaus, D.S. Boukat, K. Brander, B. Ernande, A.G. Gardmark, F. Johnston, S. Matsumura, H. Pardoe, K. Raab, A. Silva, A. Vainikka, U. Diechmann, M. Heino, A.D. Rignsdorp, "Managing Evolving Fish Stocks," *Science* 318:1247–1248 (November 23, 2007).

31 Aggie M. Blandford, "The Western Alaska Community Development Quota Program: Supporting the Advancement of Bering Sea Communities," in *Fishing People of the North: Cultures, Economies, and Management Responding to Change*, C.L. Carothers, K.R. Criddle, C.P. Chambers, P.J. Cullenberg, J.A. Fall, A.H. Himes-Cornell, J.P. Johnsen, N.S. Kimball, C.R. Menzies, E.S. Springer, eds. (Fairbanks, AK: Alaska Sea Grant, University of Alaska, Fairbanks, 2013), 243–252.

32 www.fakr.noaa.gov/sustainablefisheries/bycatch/default.htm.

33 See Thomas C. Brown and Earnest S. Burch, Jr., "Estimating the Economic Value of Subsistence Harvest of Wildlife in Alaska," in *Valuing Wildlife Resources in Alaska*, George L. Peterson, et al., eds. (Boulder, CO: Westview, 1992); John Duffield, "Nonmarket Valuation and the Courts: The Case of the Exxon Valdez," *Contemporary Economic Policy* XV:98–110 (October 1997); Robert J. Wolfe and Joseph Spaeder, "People and Salmon of the Yukon and Kuskokwim Drainages and Norton Sound in Alaska: Fishery Harvests, Culture Change, and Local Knowledge Systems," in *Pacific Salmon: Ecology and Management of Western Alaska's Populations*, Charles C. Krueger and Christian E. Zimmerman, eds. (Bethesda, MD: American Fisheries Society, 2009), 349–379; Robert J. Wolfe and Robert J. Walker, "Subsistence Economies in Alaska: Productivity, Geography, and Development Impacts," *Arctic Anthropology* 24(2):56–81 (1987); "Subsistence in Alaska: A Year 2010 Update," Alaska Department of Fish and Game, Division of Subsistence, Anchorage, AK. For critiques of economic valuation models in nonmarket economies, including hunter-gatherer economies, see

John Gowdy, ed., *Limited Wants, Unlimited Means: A Reader on Hunter-Gatherer Economics and the Environment* (Washington, DC: The Island Press, 1988) and Robert Snyder, Daniel Williams, and George Peterson, "Culture Loss and Sense of Place in Resource Valuation: Economics, Anthropology and Indigenous Cultures," in *Indigenous Peoples; Resource Management and Global Rights,* Svien Jenfoft, Henry Minde, and Ragnar Nilsen, eds. (Delft: Eburon Press, 2003). David Graeber, *Toward an Anthropological Theory of Value: The False Coin of Our Own Dreams* (New York, NY: Palgrave, 2001), David Mosse, *Cultivating Development: An Ethnography of Aid Policy and Practice* (London: Pluto Press, 2005) are also useful. Marshal Sahlins, *Culture and Practical Reason* (Chicago, IL: University of Chicago Press, 1976) remains the classic statement on the topic. These and related discussions are entirely ignored in the subsistence literature deriving from the Alaska Department of Fish and Game.

34 Kirk Johnson and Lee Van Der Voo, "Spoils of the Sea Elude Many in an Alaska Antipoverty Plan," *The New York Times* (June 19, 2013).

35 Scott A. Miller, "Economic Transition in Western Alaska Communities: Traditional Salmon Fishery Dependence and Emerging Groundfish Fishery Dependence," in *Fishing People of the North: Cultures, Economies, and Management Responding to Change,* C.L. Carothers, K.R. Criddle, C.P. Chambers, P.J. Cullenberg, J.A. Fall, A.H. Himes-Cornell, J.P. Johnsen, N.S. Kimball, C.R. Menzies, E.S. Springer, eds. (Fairbanks, AK: Alaska Sea Grant, University of Alaska, Fairbanks, 2013), 253–269.

36 Craig Medred, "Judge: Salmon Run's Survival Trumps Religious Rights of Alaska Native Fishermen," *Alaska Dispatch* (May 20, 2013).

37 North Pacific Fishery Management Council, Bearing Sea Chinook Salmon Bycatch Management, Volume 1, Final Environmental Impact Statement (December 2009), 260.

38 Ann Fienup-Riordan, *When Our Bad Season Comes: A Cultural Account of Subsistence Harvesting and Harvest Disruption on the Yukon Delta* (Anchorage, AK: Anthropological Association Monograph Series #1, 1986), 175 ff.

39 See also Stephen J. Langdon, "Subsistence Sockeye Salmon Production, Distribution, Exchange, and Customary Trade in Southeast Alaska," Prepared for Central Council of Tlingit and Haida Indians of Alaska, in fulfillment of US Forest Service Contract AG-01.-9-07-008, FIS 06-651; Nicole Gombay, "The Commoditization of Country Foods in Nunavik: A Comparative Assessment of Its Development, Applications, and Significance," *Arctic* 58:115–128 (2005), and Nicole Gombay, *Making a Living: Place, Food, and Economy in an Inuit Community* (Saskatoon: Purich Publishing, 2010).

40 Tim Ingold, "Notes on the Foraging Mode of Production," in *Hunter-gatherers. II Property, Power, and Ideology,* Tim Ingold, David Riches, James Woodburn, eds. (Oxford: Berg, 1988).

41 See Bathsheba Demuth, *Floating Coast: An Environmental History of the Bering Strait* (New York, NY: WW Norton, 2019), for an engaging overview. On notions of animal sentience among northern peoples, see, e.g., Waldemar Bogoras *The Chukchee.* The Jesup North Pacific Expedition. Memoir of the American Museum of Natural History, vol. 7. (Leiden: E.J. Brill, 1904–1909), Robert Brightman, *Grateful Prey: Rock Cree Human-Animal Relationships* (Berkeley, CA: University of California Press, 1993), Marc Brightman, Vanessa Grotti, and Olga Ulturgasheva, eds., *Animism in Rainforest and Tundra: Personhood, Animals, Plants and Things in Contemporary Amazonia and Siberia* (Oxford: Berghahn Books, 2012), Irving Hallowell, "Ojibwa Ontology, Behavior, and Worldview," in *Culture in History: Essays in Honor of Paul Radin,* Stanley Diamond, ed. (New York, NY: Columbia University Press, 1960), Paul Nadasdy, "The Gift of the Animal: The Ontology of Hunting and Human-Animal Sociality," *American Ethnologist* 34:25–43 (2007), Frank Speck, *Naskapi: Savage Hunters of the Labrador Peninsula* (Norman, OK: University of Oklahoma Press, 1935), Adrian Tanner, *Bringing Home Animals: Religious Ideology and*

Mode of Production of the Mistassini Cree Hunters (New York, NY: St. Martin's Press, 1979), and Piers Vitebsky, *The Reindeer People: Living with Animals and Spirits in Siberia* (Great Britain: Harper Collins, 2005).

42 Catherine Moncrieff, "Traditional Ecological Knowledge and Customary Trade of Subsistence Harvest Salmon on the Yukon River," U.S. Fish and Wildlife Service, Office of Subsistence Management, Fisheries Resource Monitoring Program, Final Report, Study No. 04-265 (Anchorage, AK, 2007), 16–17.

43 Moncrieff, *Traditional Ecological Knowledge and Customary Trade*, 21–24.

44 Moncrieff, *Traditional Ecological Knowledge and Customary Trade*, 27–29.

45 Theodore M. Krieg, James A. Fall, Moly B. Chthlook, Robbin LaVine, and David Koster, "Sharing Bartering, and Cash Trade of Subsistence Resources in the Bristol Bay Area, Southwest Alaska," Alaska Department of Fish and Game, Division of Subsistence Technical Paper No 326 (Juneau, 2007), 15–21, and Table 5.

46 Krieg et al., *Sharing, Bartering and Cash Trade*, 87, 89.

47 James S. Magdanz, Sandra Tahbone, Austin Ahmasuk, David S. Koster, and Brian L. Davis, "Customary Trade and Barter in Fish in the Seward Peninsula Area, Alaska," Alaska Department of Fish and Game, Division of Subsistence, Technical Paper No. 328 (Juneau, 2007).

48 Magdanz, et al., *Customary Trade and Barter in Fish*, 62–64.

49 Magdanz, et al, *Customary Trade and Barter in Fish*, 68–71.

50 *Customary Trade: Public Comments. Federal Subsistence Board Meeting Materials* (Anchorage, AK: Office of Subsistence Management, USFWS, January 2003).

51 J.W. Austin, D. Leclair, "Botulism in the North: A Disease with Borders," *Clinical Infectious Diseases* 52:593–594 (March 2011); R.P. Fagan, J.B. McLaughlin, L.J. Castrodale, B.D. Gessner, S.A. Jenkerson, E.A. Funk, T.W. Hennessy, J.P. Middaugh, J.C. Butler, "Endemic Foodborne Botulism among Alaska Native Persons—Alaska, 1947–2007," *Clinical Infectious Diseases* 52:585–592 (March 2011). R.B. Wainwright, W.L. Heyward, J.P. Middaugh, C.L. Hatheway, A.P. Harpster, T.R. Bender, "Food-Borne Botulism in Alaska, 1947–1985: Epidemiology and Clinical Findings," *The Journal of Infectious Diseases* 157:1158–1162 (June 1988).

52 See Martin David Robards, Amy Lauren Lovecraft, "Evaluating Comanagement for Social-Ecological Fit: Indigenous Priorities and Agency Mandates for Pacific Walrus," *Policy Studies Journal* 38:257–279 (2012).

53 Robert J. Wolfe, *Customary Trade in Alaska: An Introduction*, Alaska Department of Fish and Game, Division of Subsistence, Draft Report, 2000.

54 S. Langdon, R. Worl, *Distribution and Exchange of Subsistence Resources in Alaska*, Technical Paper No. 55 (Juneau, AK: Alaska Department of Fish and Game, Divisions of Subsistence, 1981). J. Magdanz, R.J. Wolfe, *The Production and Exchange of Seal Oil in Alaska* (Juneau, AK: Alaska Department of Fish and Game, Divisions of Subsistence, 1988). R.J. Wolfe and J. Magdanz, *The Sharing, Distribution, and Exchange of Wild Resources in Alaska. A Compendium of Materials Presented to the Alaska Board of Fisheries* (Juneau, AK: Alaska Department of Fish and Game, Division of Subsistence, 1993). E.S. Burch, "Modes of Exchange in Northwest Alaska," in *Hunters and Gatherers II: Property, Power, and Ideology*, T. Ingold, D. Riches, J. Woodburn, eds. (Oxford: Berg, 1988). See also J.G. Jorgensen, *Oil Age Eskimos* (Berkeley, CA: University of California Press, 1990), B.C. Hosmer, *American Indians in the Marketplace: Persistence and Innovation among the Menominees and Metlakatlans, 1870–1920* (Lawrence, KS: University Press of Kansas, 1999), and David F. Arnold, *The Fishermen's Frontier: People and Salmon in Southeast Alaska* (Seattle, WA: University of Washington Press, 2008).

55 After a decade of using this prohibitionary language, the program slightly modified its informal message:

> The only types of cash exchanges for fish and their parts that are clearly exempt from potential prosecution by the State of Alaska for violation of the State food processing laws include: (1) exchanges of cash for fish or their parts that have been processed in accordance with the State's food safety regulations, as found in 18 AAC Chapter 34; and, (2) sales of uncut, unprocessed fish.
>
> *Subsistence Management Regulations for the Harvest of Fish and Shellfish on Federal Public Lands and Waters in Alaska,* Effective April 1, 2013–March 31, 2015

56 Marion Burros, "Trying to Solve the Botulism Mystery," *The New York Times* (April 28, 1982).

57 This phrase is attributed to former governor of Alaska, Wally Hickel. Malcom B. Roberts, *The Wit and Wisdom of Wally Hickel* (Todd Publications, 1997).

58 Richard White, *The Roots of Dependency: Subsistence, Environment, and Social Change among the Choctaws, Pawnees, and Navajos* (Lincoln, NE: University of Nebraska Press, 1988).

59 Marshall Sahlins, "What Is Anthropological Enlightenment? Some Lessons of the Twentieth Century," in *Culture in Practice: Selected Essays* (Cambridge: Zone Books, 2000), 508.

60 *Transcripts of the Joint Eastern and Western Interior Alaska Subsistence Regional Advisory Council Meeting, February 24–26, 2010 in Fairbanks, Alaska* (Anchorage, AK: Office of Subsistence Management, U.S. Fish and Wildlife Service, 2010), 150–151.

5

MANAGING NATURAL RESOURCES IN ALASKA

Anthropology Bureaucratized

At a public meeting in 2011 held in Barrow, Alaska, two Iñupiat hunters described their relationships with animals, their kin, and a wider human community that included federal and state bureaucrats. James Nageak related the following:

> As—as a closing comment from me, I think that the elders that we grew up hearing and—and how we were brought up makes it hard for sometimes the elders to participate. Because we don't talk controversy about our animals, you know. We—we don't argue about them. They are part of our family. And—and when—like caribou, you know, they know when you are treating them good. I just saw a book by Harry Brower, you know. Whales, they give themselves. It's not just the whales, I think. I think that the animals we eat, the animals that we get our uses out of are very perceptive as to how we are as Inupiat people. And when there is a controversy, an argument going on that the Fish and Game or Fish and Wildlife people bring to the Inupiat people, sometimes it's hard to respond because they [the Inupiat people] don't want to get into a conflict concerning those resources that we use in our communities. And—and there could be consequences. If such a confrontation happens between let's say the Federal Wildlife Services and the Inupiat people, that we sometimes say we can't argue with you guys because the caribou are listening, the walrus are listening, the bears are listening, as to what you are saying. And so it's—it's Catch-22 for us. To—to get into a situation where we have to argue about the resources we have.[1]

Lloyd Leavitt, in his closing comments at the meeting, related a story about hunting whales with his brothers. Whales, he said, listen and respond to human discord.

> After James had said that animals do listen to us, here's a very good one. 2002, we were out—out on the ice. I was leading my brothers. And I told them

DOI: 10.4324/9781003297444-7

that we we're going to—within an hour we're going to catch the first whale. I have that feeling. Guess what? The whale did pop up, but none of my crew members wanted to catch that whale. He followed me walking. But 100 feet away he followed me right to the boat. Before that my brother was—he was in a very bad mood. And the whale just kept popping up and down. And I figured it out. I told my brother, younger brother, you need to go home for the day. Whales know when somebody's mad. They know when we're arguing. They know when the crew's not the same. We've been told that year after year. I, as a young man, pre-teen whaling, learning the ways of whaling, this whale gave himself to us. And he wouldn't—he wouldn't just give himself to us. He kept going under every time we catch—tried to get the harpoon in. Billy was in a very bad mood so I directed him to go home. I didn't ask him. You need to go home Billy. That whale is giving himself up to us and you need to go home for the day…. No more than ten minutes after he left, the snow machine sound was gone and I go that whale that—there he is. He swam about—he was about 150 feet away in front of me northwest. He swam all the way, didn't go underwater, and stopped right there, right in front of me….I believe that all creatures really do give themselves away. That is our belief and has always been our belief. And it is given to us by our ancestors. So I really believe in animals. Creatures really do give themselves if you believe in the Lord Jesus Christ.[2]

Surprisingly inflected with Christianity, such beliefs, which are not unique to the Iñupiat, present two fundamental challenges to managers of natural resources. The first challenge is conceptual. How can Western natural resource managers understand cultural beliefs that are incompatible with their own? The second challenge involves incorporating non-Western beliefs about natural and human worlds into management regimes that are driven by policy, science, and economics-based systems of knowledge.

Attempting in the late-1970s to meet these challenges, state and federal agencies developed regulations that allowed for "customary and traditional" uses of natural resources in Alaska.[3] Anthropologists and other social scientists were asked to provide definitions of and conduct research on hunting and gathering cultures throughout the state, in large part to facilitate natural resource management. Yet despite good intentions, anthropological determinations of native customs and traditions developed over the last forty years emphasize function not culture. The ensuing regulatory language reads as if written by a Western economist steeped in a utilitarian ethos.

Western ideology is everywhere apparent; native conceptions of the world are, by contrast, poorly represented. The bureaucratic language, enshrined in state and federal regulations, speaks of "factors," "use patterns," and "efficiency," as if the meaningful lives of the people in question are quite beside the point. "Custom" and "tradition" transform into ciphers for "function." Nothing resembling culture or systems of meaning is evident nor even hinted at. Iñupiat and other native understandings of the natural world and human involvement with it have been thoroughly filtered through what James Scott calls the "administrative grid" of the modern state.

Behind the bureaucratic lack of interest in symbolic systems is an associated lack of interest in the reasons people engage in their culturally situated practices. The lack of interest in both cases has the same source. Utilitarian behavior in Western economic conception is self-evident. It even appears in those who do not share the concept. All humans are thought to pursue their own self-interest in rational and measurable ways. In both state and federal bureaucracies, "pattern of use" provides the metric for native customs and traditions, and that is the end of the matter. Such is the claim and the interpretive practice. Unfortunately, for those who work within agencies managing natural resources, the maneuver precludes any serious inquiry into the ways different people meaningfully articulate their own lives, at the same time it forestalls careful historical analysis.[4]

It may be odd to discover that the bureaucratic language of custom and tradition came not from economists but from anthropologists and sociologists. These social scientists, employed by the state, were asked at an Alaska Board of Fisheries meeting in 1980 to define "customary and traditional use." They obliged, and their back-of-the-napkin, ten-factor definition, sketched out over lunch, later reduced to an eight-factor definition, has been used ever since.[5] Four decades later, contemporary state and federal anthropologists engaged in regulatory analyses of hunting and gathering in Alaska are required by regulation to use the same eight-factor back-of-the-napkin test to determine customary and traditional practices.

Over this time, something strange has happened, or rather, what is strange is that nothing has happened. Nearly four decades of anthropological research have been precluded from modifying the back-of-the-napkin definition. In 1980, culture was reduced to utilitarian behavior, custom became fetishized utility, and symbolic systems disappeared from the work of applied anthropologists engaged in influencing regulations. By 2012, the year I first drafted this discussion, and indeed today, utilitarian behavior remained at the core of anthropological research supported by state and federal agencies in Alaska, and cultural, symbolic and historical analyses were almost completely absent.[6]

The effect over these forty years has been to reinvent native culture as Western culture in disguise. Contemporary hunter-gatherers—Iñupiat, Yup'ik, Athabascan, Tlingit, Aleut, Alutiiq, Haida, Tsimshian—were thought to go about their days maximizing behavior and estimating cost/benefit ratios, engaging in customary and traditional patterns of use that were functional and pragmatic, efficient and economical. Such patterns of native use were not primarily meaningful and practiced because of those meanings and were in no way part of complex symbolic structures, their own cultures. For bureaucracies and the anthropologists they employed, counting things, including patterned behaviors, was the key to natural resource regulation, the royal road to bureaucratic enlightenment.[7]

While anthropologists became constrained by their bureaucratic blinders, local peoples began to keep count themselves, countering the state's utilitarian interests.

The Alaska Federation of Natives complained that the state's "management approach...determines the *quantitative* amount of fish and game" and ignores the "*qualitative* aspects of a subsistence way of life."[8] Such complaints were only partially effective, as native peoples were forced to accommodate to regulatory constraints and the assumptions animating those constraints, which over time altered their own cultures. The result was not simply a dominant social form imposing its will on a weaker form. Instead, a common world was constructed by diverse peoples across Alaska, but in terms of different cultural positions, some of which were mutually intelligible, and others of which were not.

One problem for applied anthropology was that functional categories of "custom" and "tradition" disallowed adequate anthropological research into the social and symbolic processes involved in the historical development of a common world. To make matters more complex, local peoples continued to express the uniqueness and integrity of their own cultures quite apart from the larger, historically contingent commonalities.[9]

A second, related problem involved the received, tacitly official history of native customary and traditional use. Passed on in institutional memory and in an impressive record of internal bureaucratic documents, this history was also uncritically replicated in a handful of scholarly publications. As it turns out, however, the history of customary and traditional use is more complex and nuanced than indicated in these various, and variously official, discourses.[10]

The original motivation for the customary and traditional use criteria was, in fact, oppositional—purposefully constructed in opposition to the Alaska boards of game and fisheries, which in the late-1970s and early-1980s were intent on undermining native hunting and gathering practices. The newly organized State Subsistence Section, under the direction of Thomas Lonner, developed criteria of subsistence use which the boards of game and fisheries could understand. Lonner himself was the author of the customary and traditional use criteria, outlined below. Couched in Western economic terms, they were intended to help preserve native culture. They stood as a proxy for native culture in a system that could not recognize the importance of native culture. As initially conceived, these criteria differentiated a Western market economy from a non-Western subsistence economy and demonstrated the persistence, internal coherence, and cultural significance of the latter.[11]

As we shall see, the oppositional motivation disappeared and over time anthropological analyses became more closely aligned with the needs of the state and less closely aligned with the needs of Alaska natives—which is particularly ironic, since early subsistence division research had a Marxist orientation.[12] Anthropologists are usually adept at analyzing the consequences of powerful institutions and their knowledge systems that "constantly organize attention away from the contradictions and contingencies of practice and the plurality of perspectives," as anthropologist David Mosse observes in his 2005 Malinowski Memorial Lecture. Indeed, in Alaska, those institutions and knowledge systems have tended to deflect any ethnographic understanding of the plurality of perspectives. They also marginalized anthropologists who highlighted the contradictions of those selfsame institutions and systems. Mosse emphasizes the need for anthropologists who work within and study powerful institutions "to negotiate space for their involvement to be more ethnographic and [to] resist institutional pressure to conform to dominant policy-driven or economics-based knowledge systems."[13] As this chapter reveals, once anthropological assessments become part of regulations governing resource use, such negotiations for conceptual space, and any attempts to resist dominant knowledge systems, appear to be foreclosed. Thus in Alaska, the institutional need to maintain authoritative—and radically simplified—representations of native customary and traditional use trumped the need for additional ethnographic research that may have called such representations into question.

Once firmly in place in resource regulations, anthropological assessments, even if deeply flawed, became difficult to dislodge. Iñupiat hunters like James Nageak and Lloyd Leavitt continued to describe their worlds, but state-employed anthropologists increasingly lost the ability to listen to them. In the bureaucratically mediated interaction between knowledge systems, Western conceptions typically held sway, but not entirely, and not without ongoing symbolic tussling.

State Involvement

When the State of Alaska was granted statehood in 1959, no one knew how to accommodate aboriginal title or even understood that it existed. However, the Alaska Statehood Act required the state to

> forever disclaim all right and title…to any lands or other property (including fishing rights), the right or title to which may be held by any Indians, Eskimos, or Aleuts (hereinafter called natives) or is held by the United States in trust for said natives….[14]

At the time, out of a total population of 226,167, there were some 43,000 Indians, Eskimos, and Aleuts in Alaska, most of whom lived in rural areas.[15] A few of them began to mount court challenges to state and federal authority over native land claims and subsistence rights, in large measure because the state government had started to claim native lands as its own, but also because of ill-conceived projects to dam a middle section of the Yukon River and to detonate atomic bombs near Point Hope in northwest Alaska as part of the Atomic Energy Commission's effort to find peaceful purposes for nuclear explosives. In both cases, the intention was to spur economic growth. The dam would provide a source of hydroelectric power far in excess of that needed by the state. The nuclear blast would dig a deep-water harbor to provide an easy passage to distant markets for coal mined in the Brooks Range. Unresolved land claims, potential large-scale environmental disruptions, and the specter of atomic explosions prompted Alaska natives to greater political activity.[16]

In response to overreaching state land claims, uncertain native claims, and heightened native political involvement, Interior Secretary Stewart Udall halted all state land selections in 1969.[17] The 1968 discovery of large reserves of oil beneath

Prudhoe Bay lent some urgency to resolving remaining state and native land selections. Without clear ownership to land, the construction of the 800-mile oil pipeline through interior Alaska would risk delay or, in the event it was built, result in prolonged litigation over claims of trespass. The political solution was the 1971 Alaska Native Claims Settlement Act (ANCSA), which effectively abolished native hunting and fishing rights, extinguished aboriginal title (to as much as 365 million acres of land), and substituted corporation for tribal ownership. "In a twinkling, but not without stunning complexity," write legal scholars David Case and David Voluck, "ANCSA converted the communal, aboriginal claims of the Alaska Natives into individual private property, represented by shares of stock in more than 200 Native regional, village, urban, and group corporations."[18] Under this Act, native corporations received forty-five million acres in land; $962.5 million in cash compensation went into the Alaska Native Fund. The state had already been allotted 103 million acres under the Statehood Act, while the federal government reserved for itself 217 million acres, some 60 percent of the state, much of which would eventually become vast parks and wildlife refuges. In this manner, the lands of Alaska were divided.[19]

Congress intended the Alaska Native Claims Settlement Act to clarify a variety of land claims issues. But it left unresolved how to administer hunting, fishing, and other subsistence activities of native peoples. State of Alaska regulations defined "subsistence" as "personal use of fish and wildlife resources," founded "on the principle of maximum sustainable yield and allocation in the broad public interest." Whenever a biological concern appeared, for example the decline in fish population below a certain threshold, then "restrictions were placed on the more efficient methods and means of harvesting personal use fish (i.e. nets, traps, snagging) … [and] a more inefficient harvest method which better served the broad public interest" was imposed.[20] State regulations, however, did not recognize or accommodate native hunting and fishing practices and categorized all of them as "personal use" subject to the transcendent principle of maximum sustainable yield. Abolished by Congress in exchange for land, native hunting and fishing rights were redefined by the state. The state, however, had little information with which to make such a redefinition and even less expertise in regulating native customary and traditional practices.

Congress had been debating native hunting and gathering on federal lands in Alaska since 1976 and would continue to debate the issue until the 1980 passage of the Alaska National Interest Lands Conservation Act (ANILCA). Among other things, this act granted the State of Alaska the right to administer hunting and fishing practices on federal lands, provided the provisions in ANILCA were met. The congressional effort over these years generated forty-two volumes of legislative history detailing the often contentious debates over native use of federal lands and the wildlife that inhabited them.[21] Part of the initial congressional intent, removed from the final act as a political compromise, was to ensure that native people had continued access to traditional resources. Remarking on the Senate bill that eventually became ANILCA, Congressman Morris Udall observed in 1980 that

At that time [the 95th Congress] we promised that any legislation enacted into law would recognize the importance of the subsistence way of life to the survival of the Alaska Native people, and would contain management provisions which recognized the responsibility of the Federal government to protect the opportunity from generation to generation for the continuation of subsistence uses by the Alaska Native people so that Alaska Native people now engaged in subsistence uses, their descendants, and their descendants' descendants, will have the opportunity to determine for themselves their own cultural orientation and the rate and degree of evolution, if any, of their Alaska Native culture.[22]

In the interim, however, Alaska passed its own subsistence statute in 1978, attempting to forestall federal involvement and retain as much control over fish and wildlife as possible. North Slope oil development, the passage of ANCSA, the debate over ANILCA, and a variety of other circumstances influenced the need for a state subsistence statute.[23] The state statute which emerged from this context defined subsistence uses of natural resources, developed a ranking of such uses (with subsistence given top priority and sport and commercial uses lower priority), and established a subsistence research program within the Department of Fish and Game. It defined subsistence as

the customary and traditional use in Alaska of wild, renewable resources for direct personal or family consumption as food, shelter, fuel, clothing, tools, or transportation; for the making and selling of handicraft articles out of non-edible by-products of fish and wildlife resources taken for personal or family consumption; and for the customary trade, barter or sharing for personal or family consumption.

The statute further defined "family" as "all persons related by blood, marriage, or adoption, and any person living within the household on a permanent basis."[24]

Customary and Traditional Use

What was "customary and traditional use"? How was it different from other non-customary and non-traditional uses? How was it to be identified? Based on what criteria? The state statute remained silent on these questions. The statute, however, established *who* would make these determinations and subsequent subsistence regulations: the board of fisheries and the board of game. Members of these boards comprised individuals appointed for three-year terms by the governor and confirmed by the legislature. Case and Voluck cite a 1978 description of these boards:

Fish and game management in Alaska is controlled by a seven-member board of fish and board of game. Both boards are dominated by white, urban

Alaskans with little allegiance to and slight knowledge of the subsistence way of life. They also have no responsibility for the overall socioeconomic effects of their policies.

As Case and Voluck go on to observe, "The boards of fish and game make wildlife management policies in splendid isolation from the rural (predominately Native) populations, which are most heavily affected by these policies."[25]

After the 1978 passage of the subsistence statute, state fish and game policies required deciding questions of custom and tradition. The results were not wildlife management policies based on scientific information, but human management policies based on questionable assumptions about human culture, behavior, and motivation. Splendid isolation allowed such boards the illusion of understanding, even as they implemented policies affecting native peoples' lives. Isolated, inexpert, focused on maximum sustainable yields, the boards of fisheries and game had the power to regulate native subsistence practices and alter native cultures. To this end, some board members "suggested that one function of the boards is to modernize rural residents, even if these residents had no ambition to be modernized," as Lonner noted in 1981. In addition, a clear bias against subsistence was manifest by the boards of fisheries and game, based on a misunderstanding of native culture.

> Many board members cannot conceive why rural residents would prefer harvesting crabs by fishing with lines through the ice, when they could be commercially profiting by owning and operating sizeable crab-fishing vessels, or why they should harvest salmon for personal consumption when they could be operating lodges catering to sport fishermen pursuing these same salmon.[26]

But the administrative reality of new subsistence regulations was more complex than ill-informed board members knew, or Case and Voluck allow, at least initially. In a 1976 policy statement on subsistence, the Alaska Board of Fish and Game and the Commissioner of the Alaska Department of Fish and Game invoked culture as a significant aspect of all subsistence fish and game management decisions. "Beyond directly satisfying food requirements," they wrote,

> home consumption of fish and game tends to preserve indigenous cultures and traditions and gives justification and gratification to a strong desire possessed by many to hunt and fish. The latter functions seem genuinely important to the physical and psychological well-being of a large number of Alaskans.

The board and commissioner further recognized that "the existing variety of cultures and life styles in Alaska are of great value and should be preserved." They advocated allocating fish and game on the basis of need, with "culture, custom, economic status, alternative resources, location, and voluntary choice of life style" as elements to consider.[27] Three years later, partially in response to the 1978 subsistence statute and partially in response to congressional debates leading to ANILCA, the boards of fisheries and game began to alter their position. Culture was still important, but "direct" dependence and "actual" need received greater emphasis. Calories, rather than culture, provided the test. Sport and commercial hunting and fishing were to be allowed alongside any priority for traditional and customary subsistence practices, providing that actual needs of subsistence users were met. The boards continued to invoke culture but in finite terms of location, traditions, customs, and alternative resources. A caloric needs test was beginning to take hold as a design tool for subsistence programs, with the boards of fisheries and game positioned to perform the requisite measures.[28]

Supporting the caloric needs test was the opinion of the Alaska Department of Fish and Game and the Department of Law. They relied on dictionary definitions of "custom" and "tradition" in order to encourage the boards to adopt regulatory language governing resource use. *Webster's Seventh New Collegiate Dictionary* was the authority. Equivalent native cultural definitions were not in play. Anthropological sources were not consulted. Based on *Webster's* authority, state biologists and attorneys devised their own characterizations of custom and tradition and arrayed them along a continuum of use. Characteristic of the continuum were oppositions between communal vs. personal, kinship vs. individual, long-term vs. short-term, and rural vs. urban. These oppositions, not the gradations between them, were the objects of regulatory interest. The master opposition, unacknowledged in the caloric test, was between native and Western culture.[29]

Also unacknowledged was a more pervasive cultural force that had the effect of "assimilating the totality of one society to the divisions of another."[30] In an attempt to regulate native culture, the boards of fisheries and game supplied their own categories of persons, environments, and economies. They had no other conceptual tools upon which to rely. These categories revealed a range of cultural values, Western in origin and orientation. Prominent among them was a clear distinction between culture and nature. By contrast, native concepts of hunting and gathering, of persons and environments, of animal motivations and the power of spoken language, remained obscure and did not form the basis of discussion. Fishing for crabs through the ice was "primitive," while crab fishing with large vessels was "modern."[31] As articulated by the boards of fisheries and game, Western concepts provided the only foundation for decisions about fish and wildlife.

Struggling to find simple language to describe a complex set of cultural practices, the State Departments of Fish and Game and Law argued that subsistence "use… may be best considered as a customary and traditional form of economic life," within which "customary and traditional use, uses, and users are inseparable from one another." This appears reasonable enough. A use requires a user. And both use and user are influenced by, indeed make sense relative to, culture. Despite the inseparability of use and user, the departments promptly divided them into separate tiers, "[f]or sake of convenience." Tier I referred to "the priority for subsistence when allocations must be made among beneficial <u>uses</u>." These uses included food, shelter, transportation, barter, and so on. In other words, the boards would decide which uses were more important for native peoples, and which were less important. Tier II referred to "the factors the Boards must, at a minimum, consider when allowable harvest is insufficient for all the <u>users</u>..."[32] In this context, the boards would decide which communities would have access to particular fish and game, and which would not.

Separating the inseparable allowed the boards a mechanism to further solidify their understanding of custom and tradition, without ever facing difficult issues of cultural exegesis. If use was about economics, and if subsistence economies were about calories, in times of shortages those calories would have to be apportioned to those most in need. In this way, customary and traditional uses provided the means of dividing up limited numbers of fish and game. How would the apportionment be made? First, by evaluating "the role of [a] resource in meeting nutritional needs." The measure for evaluation may be "the number of meals per week in which the resource is normally used." Or it may include other considerations, such as nutritional value or timing of consumption. The point was bureaucratic: to make the boards' decision making convenient, that is, "more routine."[33]

It remained unclear, however, which health experts the boards would rely upon to determine nutritional needs of natives. In any event, the boards did not address the available literature on hunter-gatherer nutrition, which would have altered any simple understanding of caloric measures, numbers of meals per week, or similar crude metrics of nourishment. John Speth and Katherine Spielmann pointed out in 1983, for example, that high levels of protein consumption as the main source of calories could lead to a variety of nutritional deficiencies among hunter-gatherers, especially if the protein source was lean meat. Sources of fat or carbohydrates, or both, would need to augment the consumption of lean meat. Ungulates hunted in late winter and early spring typically have reduced fat content. Active hunter-gatherer men required as much as 7.9 pounds of lean meat per day to meet caloric needs. Diets consisting primarily of lean meat, however, resulted in a number of deleterious physiological effects, including increased metabolism, essential fatty acid deficiency, and inhibited calcium absorption. Relying on lean meat had the effect of *raising* caloric requirements. Thus, for the boards of fisheries and game, numbers of meals per week that comprised, say, fall caribou did not provide adequate information on the role of early spring caribou in meeting nutritional needs. The boards should have had available studies before them as they determined nutritional standards of natives and other rural peoples. They did not.[34]

The shift to "actual" need was strategic. It gave the boards a rationale to accommodate sport and commercial interests and to ensure those interests were met despite the statutory mandate to do otherwise. The phrase the boards came to rely upon was "amounts necessary for subsistence."[35] With that phrase, culture completely dropped away. Starting in the early 1980s, applied anthropologists employed by the state expended considerable effort to count fish and game used by rural peoples in order to provide numbers for the boards to consider. Local peoples, however, did not determine their own necessary amounts based on their own meaningful practices.

The boards of fisheries and game made those determinations, mostly in an attempt to limit subsistence to "amounts necessary." With limits to subsistence, the boards could promote sport hunting and fishing as well as commercial uses of fish and game. The process was open to continual contestation and caused rural peoples to distrust board determinations.[36] Still, rural peoples, native and non-native, had to adjust their interests to those of more powerful players, whose decisions mattered more than their own.

As with calories, so with Western economies. Each is an abstract system of value undergirded by notions of scarcity and a presumed universal human drive for satisfying personal wants and needs. Each relies on the supposition that humans rationally pursue a maximization of their needs. Calories in one system, merchandise in the other. From this point of view, calories from salmon are equivalent to those from bowhead whale, just as cash is equivalent whether used to purchase a rod and reel or an airline ticket to Hawaii. The question in both cases is how to make use of limited means. Thus, as Sahlins frames the general economic issue, "things unlike in their objective attributes and human virtues—their different meanings to us as use-values—are indeed comparable as exchange-values."[37] Such comparability comes at a cost, however. The specific virtues humans ascribe to whales and salmon disappeared when the exchange-value was caloric. In this way, the boards could dismiss culture as peripheral to the central problem. Despite several years describing the importance of culture, when it came time to allocate scarce resources the boards of fisheries and game fell back on familiar territory—familiar because the same logic that animated their economy was easy to transpose onto allocations of scare game. A cost/benefit analysis would suffice in either case. No need to understand culturally mediated use-values. Calories were calories. Custom and tradition had come to that.[38]

Meaningful Orders of Persons and Things

Use-values—culturally specific valuations of the natural world—are implausibly transformed into exchange-values. But transformed they were in a transmutation as magical and conventional as transforming for ritual consumption stale bread into flesh. By reducing cultural characteristics to market values or, as with calories, displaced market values, the boards of fisheries and game were well on their way to mastering local cultural orders. Scarcity (calories) and need (customary and traditional use) replicated conventional economic wisdom, not the wisdom of native peoples. Armed with their own cultural presuppositions about the universality of human desires and actions and fortified by their beliefs about the differences between humans and other animals, the boards, however, could not entirely master local cultural orders.

Natives, meanwhile, continued to assimilate regulations into their own cultural schemes. According to Mr. Nageak, with whose comments this chapter began, controversies between Iñupiat and state and federal agencies must remain quiet for fear that the animals, sentient creatures in their own right, would hear and understand and respond.[39] Similarly for Yup'ik Eskimo, among whom great care must be taken when interacting with animals. For Yup'ik, all creatures are sentient non-human persons endowed with understanding and a desire and ability to control their own destinies. A Yup'ik from Scammon Bay, Alaska, remarked, "All the animals or birds decrease or increase when they want to. Although many are killed, they increase again. Geese don't decrease to extinction: they become plentiful when they want to." Some Yup'ik believe that the initial decline of geese on the southwest coast of Alaska was the result of improper activities, as Western biologists handled nests, eggs, and goslings during their bird studies. But ultimately geese decide if and when they will return. Another Yup'ik put it this way:

> Sometimes we can't really blame anything and anyone for the decrease of geese. When they want to be fewer within a year they are fewer. We don't know why this happens, but the next year there are more. Even though a lot are killed, their numbers don't seem to change the following year.[40]

Inanimate objects, as anthropologist Julie Cruikshank has shown, also have their own sentience—glaciers, for example, which can hear and smell and act—at least those glaciers in Tlingit and Athabascan conception. Human-like sensory perception is accompanied by a kind of moral authority. Glaciers are believed to actively punish those who transgress certain norms. Cooked food is particularly noxious, and glaciers can smell grease cooking on a fire. A glacier may respond by a rapid surge, obliterating a town on its margins, or by a surge that dams a river and forms a new lake. Tlingit and Athabascan, who live in the Mount Saint Elias area of Alaska and Canada, don't know exactly why this happens but insist that it results from human folly, from transgressions of the moral order.[41]

Among Koyukon Athabascans in Huslia, Alaska, fire is thought to be sentient. Since the spoken word has power, one has to refrain from speaking ill of fire or indeed of other natural forces, "lest they hear and take offence," as Henry Huntington and colleagues have noted. "One elder told us of how she had been instructed not to speak of what a fire might do because speaking in that way might cause the fire to do exactly as she had said."[42] Similarly, talk of climate change may result in the very change one discusses—and so one must avoid speaking publicly of the weather or other "big things."[43]

The world is very different if it is filled with animate and inanimate non-human persons who influence one another and can bestow good fortune upon, or cause harm to befall, humanity. How can Western scientists and bureaucrats understand and incorporate such ideas into subsistence management practices? What happens when beliefs that cannot be empirically verified clash with Western notions of environments and the creatures that inhabit them? Cruikshank is not very hopeful.

She notes, citing a variety of anthropological studies, that natural resource managers use cultural "tradition" insofar as it reflects "ideas compatible with state adminis-tration rather than those understood by local people."[44] Cultural "traditions" that are incompatible with administration and the cultural values which animate it are generally ignored. Hence, there is the reduction of native customs and traditions to calories based on the familiar logic of cost/benefit relations. Western customs and traditions are thereby maintained.

Cultural incompatibility, however, is exacerbated by an anthropological focus on "economic" aspects of subsistence, which replicate Western categories of use. Western culture has long posited a nature/human dichotomy, easily expressed in economic terms and easily overlooked as a product of Western culture. This dichotomy takes a variety of forms in Western thought, including the insistence on the uniqueness of humans as distinct from other animals. One consequence, as anthologist Tim Ingold observes, is that "we can countenance an enquiry into the animal nature of human beings while rejecting out of hand the possibility of an enquiry into the humanity of non-human animals....Humans are both persons *and* organisms, animals are all organism."[45]

Iñupiat, Yup'ik, and Athabascan peoples in Alaska reject this view, as do many other northern peoples. For example, Cree notions of personhood encompass humans, non-human animals, and other "natural" phenomena:

> In the culturally constructed world of the Waswanipi [Cree] the animals, the winds and many other phenomena are thought of as being 'like persons' in that they act intelligently and have wills and idiosyncrasies, and understand and are understood by men. Causality, therefore, is personal not mechanical or biological, and it is...always appropriate to ask "who did it?" and "why?" rather than "how does that work?"[46]

One aspect of understanding the world as filled with intelligent non-human beings is that the environment and its inhabitants are not seen as distinct from humanity:

> The point is not that the difference between (say) a goose and a man is between an organism and a person, but between one kind of organism-person and another. From the Cree perspective, personhood is not the manifest form of humanity; rather the human is one of many outward forms of personhood. And so when Cree hunters claim that a goose is in some sense like a man, or that the two are even consubstantial, far from drawing a figurative parallel across two fundamentally separate domains, they are rather pointing to the real unity that underwrites their differentiation.[47]

Division of Subsistence

It was in the context of calories and culture, and the discordance between very different systems of knowledge, that the boards of fisheries and game turned to social scientists for advice. Not by choice. The state subsistence statute also established the subsistence section, a research arm of the Department of Fish and Game. The work of the subsistence section was met from the very beginning with "a significant amount of institutional inertia and even active resistance," as Lonner observed in a 1980 internal memorandum. The memorandum understated the problem of institutional resistance and its consequences: "It is unclear whether certain 'bodies' have the desire and capacity to incorporate new socioeconomic and sociocultural information."[48] Among these bodies were the boards of fisheries and game, which have continued into the twenty-first century to resist adequate anthropological research. Their preference was (and remains) for easy-to-digest worksheets indicating customary and traditional uses, and simple surveys indicating numbers of animals killed for subsistence along with associated estimates of food value. The practices became institutionally secure enough to eventually have their own initialisms: C&T and ANS.[49]

Institutional opposition, however, was not limited to the boards of fisheries and game. The Department of Fish and Game drafted a 1982 memorandum from 35 "unnamed Department signatories to the Commissioner of Fish and Game," asserting that the subsistence law resulted in a wide-ranging and uncontrollable congeries of ills. They alleged that it caused the "dissipation of department resources, management by the courts, fragmentation of the regulatory process, alteration of data collection, increased enforcement problems, [and] shrinking constituency for fish and wildlife." Perhaps most damaging of all, certainly most damning, the subsistence law led to "the politicization of the Department's Subsistence Division, through the introduction of social science."[50] For the Department of Fish and Game, danger was lurking in the guise of the human sciences. Funds were diverted, regulations were made more complex, and biologists had to contend with alien notions of the natural world. Research on humans had the potential to undermine assumptions about natural resource management, and the signatories of this memorandum felt threatened.

Although the biologists' fears were never entirely allayed, the subsistence section of the Department of Fish and Game ultimately accommodated the boards' biological preferences, but at the expense of isolating itself from wider, perhaps even more dangerous, intellectual trends in anthropology. At its inception, however, the subsistence section embarked on a research program which was not limited to C&T and ANS. In 1979, Lonner wrote that the role of the subsistence section was to provide the boards with research-based evidence with which they could rationally define and regulate subsistence. Until such evidence was made available, the boards should refrain from defining "subsistence." Why? "The practice of subsistence has a 10,000 year history in Alaska and the English word by which it is known may be a very inappropriate label for the activity which continues into the twentieth century."[51] Lonner believed the boards needed to wait for the subsistence section to develop adequate research in order to avoid making ill-informed regulatory choices based on incomplete or erroneous information. The boards, however, could not wait for adequate research.

Criteria of Subsistence Use

In 1980, the Alaska Board of Fisheries adopted Lonner's ten criteria of customary and traditional practices, reasoning that they could be used "to identify subsistence uses."[52] In 1981, the Alaska Boards of Fisheries and Game issued a resolution, later adopted into state regulation, which specified how subsistence uses would be described and managed. The boards came up with the following elements, slightly modifying Lonner's original ten criteria. "Customary" and "traditional" subsistence uses were identified by the boards as exhibiting the following:

1. a long-term, consistent pattern of use (excluding interruption by circumstances beyond the users' control such as regulatory prohibitions);
2. a use pattern recurring in specific seasons of each year;
3. a use pattern consisting of methods and means of harvest which are characterized by efficiency and economy of effort and cost and conditioned by local circumstances;
4. the consistent harvest and use of fish or game, which is near or reasonably accessible from the user's residence;
5. the means of handling, preparing, preserving, and storing fish and game, which has been traditionally used by past generations (but not excluding recent technological advances where appropriate);
6. a use pattern, which includes the handing down of knowledge of fishing or hunting skills, values, and lore from generation to generation;
7. a use pattern in which the hunting or fishing effort or the products of that effort are distributed or shared among others within a definable "community" of persons, including customary trade, barter, sharing, and gift-giving. Customary trade may include limited exchanges for cash but does not include significant commercial enterprises. A "community" for purposes of subsistence uses may include specific villages or towns, with a historical preponderance of subsistence users and in addition encompasses individuals, families, or groups who in fact meet the criteria described in this policy;
8. a use pattern which includes reliance for subsistence purposes upon a wide diversity of the fish and game resources of an area, and in which that pattern of subsistence use provides substantial economic, cultural, social, and nutritional elements of the subsistence user's life.[53]

The state boards of fisheries and game, and, starting in 1990, the Federal Subsistence Board used this list to make species-by-species determinations of customary and traditional use.[54] By 2012, the number of customary and traditional use determinations throughout Alaska had exceeded 300, each founded on an anthropological analysis of patterns of resource use. It is here that the work of applied anthropologists and the administrative grid of the modern state became fully integrated. It is here that the normalizing process of bureaucratic effort became manifest. And it is here that function, not culture, animated the work of bureaucrats, even as they talked in terms of "traditions" and "customs."

Yet by 2012, as in 1980, and indeed today, the unexamined assumptions remained the same. Both state and federal programs assumed that the eight criteria of subsistence use adequately demonstrated local customs and traditions, and that, consequently, anthropological analyses based upon them were empirically reliable. Were these assumptions tenable?

The available literature contains numerous historical, ethnographic, and theoretical counterexamples to each of the eight criteria, which should have called into question their apparent commonsense applicability. The fact that one can come to quite opposite conclusions from the same sources indicates that other forces were at play, skewing subsistence information for particular purposes rather than presenting an objective record, and stabilizing a certain set of interpretations at the expense of other interpretations—again, part of the normalizing bureaucratic process.

We will examine each category against the available literature in an attempt to understand why they persist and to uncover how the administrative grid is central to such persistence. The general point I wish to emphasize is this: once locked into state and federal regulation via the usual bureaucratic processes, adequate critique of those processes becomes difficult if not impossible as the various participants—anthropologists, bureaucrats, board members, rural peoples—adopt those processes as their own. In this way, any friction between different systems of cultural knowledge is minimized. I'm interested in what is lost in such bureaucratic minimization.

Against "Customary and Traditional Use"

Long-Term Subsistence Patterns

According to the boards of fisheries and game, subsistence is characterized by a "long-term, consistent pattern of use." Yet many contemporary native patterns resulted from European contact, the intrusion of capitalist markets, and the arrival of religious emissaries into native communities. New species were hunted in support of European economies in the nineteenth century, certain subsistence practices were lost as a consequence, and in some instances natives became middlemen traders and gave up subsistence practices entirely. Criterion 1 fails to take into account any relevant historical alterations to local cultural patterns, despite the parenthetical remark about regulatory circumstances beyond users' control. "Consistent pattern of use" ignores well over a century of historical contingency and subsequent changes to native productive patterns. "Changing patterns of use" may have been a more accurate assessment.[55]

Robert McKennan explicitly argues that new trade relations in the late nineteenth century altered local cultures along the Tanana River in central Alaska. "These trade contacts," he writes, "brought inevitable changes in the material culture of the Indians, but what is more important, they profoundly affected their subsistence pattern, round of seasonal activities, social organization and demography."

> Semi-permanent villages grew up in the neighborhood of trading posts, and a market developed for fish as well as furs. The introduction of the dog team and its growing use in fur trapping further increased the need for dried fish, an

easily transportable dog food. More and more the natives' economic life came to center around the individualistic activities of the nuclear family, rather than the earlier collective activities of the larger band, itself little more than a large extended family under the leadership of a patriarchal 'chief.' Richard Slobodin has noted similar changes for the Peel River Kutchin [Gwich'in]. These early alterations were intensified by other important agencies of cultural and demographic change, here listed in chronological order: (1) steamboats, utilizing native pilots and crews, resulting in great geographic mobility for many of these Indians; (2) establishment of missions, and particularly missions schools, at Tanana (1887), Minto (1929) Nenana (1907), Chena (1908), Salcha (1909), and Tanana Crossing (1912), which drew natives from various groups to semi-permanent settlements along the river; and (3) introduction of the fishwheel, which transferred fishing activities from the clearwater streams and interior lakes to the muddy waters of the lower Tanana.[56]

James VanStone makes a similar point in reference to the Yukon River Ingalik. He writes that late-nineteenth century forces altered native productive patterns:

> Traditional subsistence activities may have been modified much earlier but in the early American period, for the first time, trapping was almost certainly pursued to the detriment of the hunting of large game animals during the fall and winter months. As a result, the quantity of meat available was greatly reduced with a corresponding increase in dependence on fish and on supplies obtained from the traders. By 1883 the Ingalik were, in fact, dependent on American commodities and the trading post was no longer simply a source of exotic luxury goods, but a necessity for survival.[57]

Shepard Krech III emphasizes the cultural changes resulting from novel trade networks. He notes that European trade strengthened some existing native trade relations and altered others, which had the effect of modifying native productive practices.

> Some bands assumed lucrative middlemen positions which they become reluctant to relinquish. The Yukon Flats Kutchin [Gwich'in] became extremely aggressive after the establishment of Fort Yukon in their territory (Anderson: 1858 report). Several years later, they were reported to "hunt no furs" (Hardisty 1872:311; cf. Kirby 1872:418) and to "make very little for themselves, but buy from other Indians" (Jones 1972:325). The Upper Porcupine River Kutchin were reported to be "enraged" at the establishment of Fort Yukon (Murray 1910:55) and at the subsequent erosion of their trading position between the Western Kutchin and the Peel River Post. They had been the most important middlemen in the trade at Peel River Post prior to 1847; by 1858 they bartered only muskrats (Peers: 3/25/1852; Anderson: 1858 Report).[58]

Richard Nelson similarly recognizes profound changes to native subsistence patterns. Writing in 1973, Nelson observes that "Fur brought the white man into Kutchin country, fur kept him there, and fur has been the nexus between the Indian and the world outside for most of the past 120 years." He goes on to say that

> The Kutchin had always taken some fur animals, using their highly effective deadfalls and snares, but for the most part these animals were peripheral to the native economy. Now the white man wanted fur and would pay handsomely for it with highly desired trade goods. That was enough for the Kutchin, who began to devote more and more effort to catching furbearers. It is hard to say when trapping became a dominant force in their lives, but by the turn of the century they were probably modifying their entire life-style to fit into a trapping regime.[59]

The modified lifestyle persisted for half a century. A decline in fur prices after World War II, an increase in summer employment, the availability of welfare payments, and the construction of schools, however, combined to encourage people to move to larger villages. "The old trapping life," Nelson writes, "which had become as much the pattern of Kutchin culture as the nomadic hunting and fishing existence which preceded it, began to disappear."[60]

Changes to productive practices came comparatively late to natives in interior Alaska. On the coast of Alaska, by contrast, change came much earlier. Dorothy Jean Ray points out that the 1789 establishment of a trading market on the Anyui tributary of the Kolyma River in Siberia had "far-reaching effects in Alaska." This market, primarily for furs and tobacco, "created the first change in the economy of the historical Eskimos," because fox and marten furs "destined for European consumption, were not traditionally used by Alaskan Eskimos except in ornamentation of fur clothing." Ray goes on to emphasize the resultant changes to subsistence patterns. "Alaskans not only began trading as links in a chain of traders that originated in Indian territory to the east and ended at Anyui, but began trapping the animals themselves."[61] Subsistence became tied to the global economy, especially as Alaska natives adopted the tobacco habit, trading furs headed to Europe for tobacco that had come from Poland or Sweden. Tobacco was traded across Siberia and eventually to Chukchi middlemen who traded it to their Alaskan counterparts for furs.[62]

By the end of the nineteenth century, concurrent with the rise and decline of the whaling industry, coastal Eskimo were affected by and eventually incorporated into that industry and then were simply abandoned by it. Initially, Eskimo traded furs to whalers for tobacco and other European goods. Natives on St. Lawrence Island, the Diomede Islands, and the Chukchi Peninsula also provided locally made boots, parkas, and mittens to outfit the whaling crews for northern latitudes, making a substantial income as a result. Later, coastal Eskimo were recruited as sailors on whaling ships, which resulted in shifts in local populations. In the 1890s, Point

Hope Eskimo moved to Point Barrow, and those from the Seward Peninsula and rivers draining into Kotzebue Sound moved to Point Hope. Such was the international and local mix of languages at the remote village of Point Hope that it earned the sobriquet "Jabbertown."[63]

In the long-term, consistent patterns of resource use were thus subject to the increasing intrusion of the global economy, beginning some 200 years before the boards adopted this category as a stable marker of subsistence. In response to these and other nineteenth and twentieth century changes, natives altered their productive practices, and variously so in different parts of Alaska. Markets for furs, whales, fish, and gold all affected native practices. Criterion 1, then, mistakes as "traditional" diverse social and cultural patterns that were, in fact, products of complex histories. This is not to suggest that there were no patterns of native resource use, nor that practical considerations of food supply and security diminished with native involvement in European markets. But, depending upon the native group and its relation to global forces, the disruption of older patterns and the adoption of new ones were more important than either duration or consistency of use.

Rather than empirically justified, the "long-term, consistent pattern of use" category served to ease the administrative burden carried by boards of fisheries and game. It provided the illusion of continuity of culture. Pointing to such a pattern, backed by limited anthropological research, the boards emphasized "use" in a way that fixed use in time and place and, thus simplified, allowed for the administration of fish and game—the true purpose of these boards.

Seasonality

The idea of seasonality suffers from similar conceptual and empirical problems. It assumes seasonality of use, but many species, caribou for instance, unexpectedly changed their migratory patterns and humans were forced to turn to other species for sustenance, simply because caribou were too distant to locate or arrived months later than anticipated. Shifting migratory patterns became an increasing concern after native peoples settled in villages in the early twentieth century. Pertinent examples are easy to find. The small groups of Nunamiut who inhabited the central Brooks Range in the late-nineteenth and early-twentieth centuries were forced to move after a crash of the Western Arctic Caribou Herd. By 1920, this area of Alaska was effectively depopulated of humans. As a consequence, there was no seasonality of use until, years later, people returned to Anaktuvuk Pass. Multiple factors were associated with the depopulation, such as a decline in trading between natives, the establishment of missions and schools on the coast, a shift from whaling to fur trapping as a primary source of cash for natives, as well as a devastating caribou herd crash.[64] Seasonality, in other words, was rarely a simple, empirically obvious aspect of long-term native resource use.

In addition, not all peoples had clear seasonal rounds. Among the Eyak in southeast Alaska, to cite Kaj Birket-Smith and Frederica de Laguna's 1938 assessment, "There was not much of a yearly cycle in hunting."[65] Seasons were important but clearly secondary to the behavior of animals, whose own seasonality was open to disruption from any number of natural causes. Changing river courses destroy timber stands and redeposit soils. Fires alter vegetation patterns, which may benefit moose but not caribou. Anadromous fish may return to specific streams in large or small numbers to spawn, based on ocean conditions and associated La Niña and El Niño events. In response to unpredictable changes, subsistence hunters minimized their harvest territories when resources were especially abundant and expanded them when they were not. And when resources were less abundant, trade relations became even more important. Flexibility was the key adaptation when seasonal patterns failed. Such flexibility extended to novel social relations as well.[66]

By the time Birket-Smith and de Laguna conducted their fieldwork in the early-twentieth century, native seasonality of resource use had been thoroughly tempered by market economies in southeast Alaska. The late-nineteenth century gold rush altered any harvest seasonality by providing substantial economic opportunities for local natives. For example, Tlingit entrepreneurs charged between 14 cents and 45 cents a pound to pack miners' gear up Chilcoot Pass in 1897. The profits were high, since thousands of stampeding miners typically brought with them a half ton or more of food and equipment. Seasons and transportation costs were linked, and Tlingit demanded less per pound in winter when sleds eased the burden of transportation, and more in summer when the physical effort was much greater. Over the first two seasons of the Klondike gold rush, natives "exerted significant control over the finances of transportation at Dyea and at the Chilkoot summit," as Kathryn Morse observes in *The Nature of Gold*.[67] For the Tlingit, the cash they earned, and the goods that could be purchased with it, had a particular purpose: to be given away. Ritualized gift-giving increased a person's status, which was more highly valued than cash or European goods. Many Tlingit thus linked their own cultural concerns with those of Westerners, played out in the physical effort of moving the goods of an industrialized society, from pickaxes to cans of beans, in support of members of an industrial society who were searching for their own profit in the form of gold. For Tlingit, however, opportunistic hunting made way for opportunistic profit making, and seasonal variations provided differential rates of monetary return, not opportunities for subsistence.

The seasonality of use criterion assumed a stable natural and social environment. However, the late-nineteenth century influx of gold seekers, as they hunted for their own needs, added additional pressure to moose, caribou, and mountain sheep, decreasing their numbers for native use. After the turn of the century, and for many decades thereafter, accidental and deliberately set fires altered habitats, and "populations of moose, caribou, mountain sheep, wolves, and bears were reduced to historically low levels." Moreover, in the mid-twentieth century, the U.S. Fish and Wildlife Service embarked on an aggressive predator control program, disrupting population dynamics of a variety of prey species—moose, caribou, deer—which once again altered their availability for human harvest by artificially increasing their numbers. In these instances, seasonality became an artifact of management and not directly related to any particular long-term pattern of native use. The patterns themselves were constantly changing.[68]

Invoking seasonality also raises the ethnographic question: whose seasons? Koyukon Athabascans, who reside along the lower and middle reaches of the Yukon River, recognized 16 seasonal transitions, with a great deal of individual variation on precise timing:

1. *Me nen tar-kenet'oihe* (the time when the sun is under water)
2. *Me mem miltsa dzan yedelaihe* (the time of lengthening days)
3. *Telel-soo* (time of the eagles)
4. *Kolekeih-soo* (time of the hawk)
5. *Kaka-soo* (time of the geese or time of the animals)
6. *Menen lu tedare* (time of the breaking of the rivers)
7. *Me nen to-tsih-leyaihe* (the time for launching canoes)
8. *Me nen keton nelyaihe* (time during which leaves grow)
9. *Kal-nora* (time of the king salmon)
10. *Nular-nora* (time of the dog salmon)
11. *Sanlar-nora* (time of the silver salmon)
12. *Noltlar-nora* (time of the second run of the dog salmon)
13. *Roihtse-luka-nora* (time of the autumn fish)
14. *Men nen tedetihe* (time of the freezing)
15. *Men nen ketsunelainhe* (time during which we trap martin)
16. *Men nen ral a-tlo-karaleyaihe* (time when marten turns its behind to trap)[69]

Rather than the category "seasonality of use," opportunistic hunting would be a more accurate category, in line with the substantial archaeological, historical, and ethnographic record on hunter-gatherers.[70] Mobility, flexibility, and alternative patterns of use were characteristic of hunter-gatherers in Alaska. Opportunistic hunting, however, would be difficult to govern. It was never seriously considered characteristic of customary and traditional use or contemporary subsistence. The boards' intent was wildlife management founded on the ever-present need for maximum sustainable yield. The boards were less interested in cultural accuracy or empirical precision. They certainly did not want to allow unpredictable, hence uncontrollable, resource use. Maximum sustainable yield came first. If there was a master category, maximum sustainable yield was it.

Efficiency

Others of the boards' criteria were similarly problematic. Criterion 3, which posits "efficiency and economy of effort and cost" as characteristic of subsistence practices, promoted a functional perspective, as if "efficiency" and "effort" were objective aspects of hunting and gathering. Yet as early as Zagoskin's account of his 1842–1844 explorations of the upper Yukon River area, we read the following about local practices on the Nulato River:

> The river is…well supplied with otter, and the natives shoot both these and the beaver with bow and arrows. A special sort of pinewood trap is set for otter in the winter, but as this type of hunting is considered easy, it is left to the shamans and the old men.[71]

For these people, economy of effort was marked as the realm of the aged; effortful hunting was the realm of the young and robust.

Efficiency and economy of effort were also relative to the value native peoples placed on particular species. Some species simply were not hunted even if minimal effort was required. The Eyak did not hunt walrus, land otters, or wolves, who were thought to be humans transformed into these forms. If the need arose, wolves would be reasoned with, not hunted, and in turn would be persuaded not to hunt humans. Birket-Smith and de Laguna write that "The Eyak believed that if attacked by a wolf they could speak to it and induce it not to hunt them."[72] In such cases, effort was directed at communication with the human/wolf, not with the procurement of food or furs.

In other cases, areas rich in game were avoided for cultural reasons. Traditional Iñupiat in the early-nineteenth century had a keen understanding of their environment. "It is difficult," Ernest Burch writes, "to see how any people relying entirely on resources found in northwestern Alaska could have improved on the Inupiat approach to subsistence." Burch goes on to observe,

> The human adjustments included highly rational techniques for locating, pursuing, striking, killing, retrieving, butchering, and storing the fish and animals on which they relied for survival. However, they also included a complex array of ceremonial, ritual, and other magical practices. In some cases, people could not even live (or hunt or fish) in rich hunting or fishing areas because of their fears of nonempirical phenomena, such as ghosts of one kind or another.[73]

As Burch points out, cultural concerns required leaving an area and its animals unmolested, even when efficiency indicated otherwise. Burch further argues that when hunting rituals and taboos failed—that is, when the hunt failed—shaman tended to add

> new taboos and rituals onto old ones rather than replacing them with something new. Over centuries, this resulted in the accumulation of a tremendous number of restrictions on people's freedom of action in pursuing, killing, processing, and consuming fish and game.[74]

Without accounting for culture, in other words, "efficiency and economy of effort and cost" reveals very little about a people's subsistence uses, and even less about customs and traditions. Efficiency and economy of effort, however, proved long lasting as a category in the boards' administrative grid of control.

Proximity

Among Yup'ik, as Fienup-Riordan shows, many valued resources take great effort to harvest. The resources with highest value are often most difficult and least efficient to obtain. In addition, with a shift to permanent villages in the early-twentieth century, older patterns of resource exploitation continued to be followed into the 1980s:

> In comparisons both within and between the villages of the lower Yukon delta and Scammon Bay, it is apparent that current village residents continue to exploit the territory of the village group in which they were born, regardless of the most proximate location of the resource or the distance from their present place of residence.

In addition, in the context of sharing among members of extended families, "diversity [of resources] rates over efficiency, and variety over maximum productivity."[75] The "efficiency and economy of effort and cost" criterion appeared to be directly contradicted by the requirements of sharing among kin.

On occasion, villagers would charter planes to gather berries from distant sources, usually in conjunction with visiting relatives. A Division of Subsistence report cites the following examples from 1983:

> When an economic arrangement involving non-local subsistence products is made, usually currency is utilized. For example, this summer some Togiak residents chartered a flight to visit relatives in Nunapitchuk located in a tundra area renowned for its abundance of salmonberries. The Togiak visitors had prearranged for their first cousins to pick several buckets of salmonberries for them. The buckets of berries were purchased, but reportedly, the money was not for the berries, but for the effort of gathering. Similarly some Quinhagak residents take the mail flight to Platinum and Goodnews Bay each fall, as these areas have more abundant blackberries than Quinhagak. In these cases, the women pick berries with relatives in the communities, bringing the berries back in plastic five-gallon containers.[76]

As a category, "proximity" appears commonsensical. Why would anyone go long distances to gather food? But in the cultural event, the category is unreliable and ignores the reasons people have for gathering foods from distant places, which gathering may only obliquely be about the food itself.

One would think the boards would understand this aspect of subsistence, since sport hunting and fishing often require long-distance travel and the expenditure of large sums of money for relatively little return. Cost-benefit analyses do not seem to be part of the effort of sport hunters and fishers. But I can't recall, in the extensive reading I've done in the transcripts of subsistence meetings, any such understanding expressed by members of either the board of game or the board of fisheries.

Accessibility

The pattern of exploiting distant resources violates criteria 3 and 4, the latter of which refers to resource harvests "near or reasonably accessible from a user's residence." Yet accessibility cannot be measured simply by geographic proximity. Accessibility is always relative to the technology used to get from place to place. Dog teams increase accessibility. Snow machines increase it further still, and airplanes even more so. Dog traction did not become important along the Yukon River until the second half of the nineteenth century, when dog teams were used in support of the developing fur trade and, later, in support of gold miners. New networks of trails emerged. New social, economic and productive practices emerged as well. By the early-twentieth century, there were some 6,000 dogs in the Yukon River drainage, used to pull sleds for transportation of people and goods and, increasingly over the next 100 years, for competition. These dogs needed to be fed. The result was a greater native emphasis on summer salmon fishing, which in turn changed other productive practices. A U.S. Bureau of Fisheries survey in 1918 estimated that one million salmon were needed annually to feed the dogs in the drainage. Despite the increasing importance of aircraft, starting in the 1930s, and the adoption of snow machines, starting in the 1970s, the numbers of dogs in villages along the Yukon River remained high. By the early 1990s, an estimated 5,000 sled dogs continued to be fed subsistence-caught salmon. Since that time, those numbers have been in sharp decline, reduced by as much as 50 percent by 2008.[77]

In addition to technological considerations, accessibility must be understood relative to one's desire and willingness to travel some distance to hunt, trap, and fish—that is, to cultural factors of travel, distance, territory, notions of time, physical exertion, the cultural value of particular species, and kinship. Traditionally, a multi-week or multi-month hunting trip was common. One had to travel to intercept migrating caribou or hunt sheep. Moreover, in hunter-gatherer economies, non-utilitarian mobility across large distances was often as important as utilitarian mobility. Maintaining networks of kin across expansive geographic regions has been recognized as central to hunter-gatherer adaptations. Unlike subsistence networks, marriage networks of hunter-gatherers were often regional and allowed movement from areas of scarce resources to areas of more abundant resources. Divorcing subsistence networks from marriage networks may be germane for wildlife management, but such a divorce effectively negates the cultural importance of regional ties and shrinks geographic space down to a bureaucratically convenient size.[78]

One need not comb the archaeological, historical, or ethnographic literature for relevant examples of long-distance travel. A 2012 Bureau of Land Management "Draft National Petroleum Reserve-Alaska Integrated Activity Plan/Environmental Impact Statement," focused on Alaska's North Slope, notes the following:

> Several times in the 1970s and 1980s and as recently as 1994 and 1998, Anaktuvuk Pass residents found it necessary to travel great distances to procure enough caribou to feed their community. The North Slope Borough paid for some trips, using charter and float planes to fly hunters from Anaktuvuk Pass to places like Umiat and Schrader Lake (located approximately 60 miles southwest of Kaktovik…).[79]

Under subsistence criteria, such activities would be neither traditional nor customary, since they involved harvesting distant caribou. Anaktuvuk Pass residents, however, would have no difficulty asserting their customs as the reason to hunt caribou, notwithstanding an extended, uneconomical airplane flight. Despite evidence to the contrary, this category has remained securely within the bureaucratic framework regulating subsistence hunting and fishing.

Preparing, Preserving, and Storing

The boards consistently undervalued the importance of culture in determining customary and traditional uses. However, in its reference to traditional means of preparing, preserving, and storing fish and game, criterion 5 appeared to address cultural issues. Yet here, too, recent technology was invoked parenthetically, and it remained unclear which past generation's use would be privileged in the event new technology altered such use. Traditionally, storage of any fish or game beyond a season or two was rare; new technology allowed much longer storage times. Jarring of salmon has become commonplace among residents who live along the Yukon River, for example. But to understand storage one must also understand reasons not to store foods for future use. Among the Koyukon Athabascans, accumulation of fish and game for future needs, rather than sharing with others for immediate needs, was open to ridicule. Accumulation was miserly, long-term storage unnecessary. Spiritual reprisals and ill luck would befall a stingy person, manifested as poor hunting success or personal injury. Best to share rather than to store.[80]

Still, the category of preparing, preserving, and storing appeared obvious, and local practices, even as they changed, needed to be described and worked into the determinations of the boards of fisheries and game, even if historical change was only of parenthetical concern.

Knowledge

The customary and traditional use pattern of criterion 6, "the handing down of knowledge of fishing or hunting skills, values, and lore from generation to generation," brings anthropologist Claude Lévi-Strauss's once well-known observations to mind. After reviewing examples of elaborate classifications and highly detailed understandings of plants and animals from diverse cultures, he notes that

> Examples like these could be drawn from all parts of the world and one may readily conclude that animals and plants are not known as a result of their usefulness; they are deemed to be useful or interesting because they are first of all known.[81]

Such a position, unacknowledged in the boards' functionalist orientation and unarticulated in subsistence division research, was prominent in the anthropology of the day. The elementary point is that understanding culture, including culturally specific classifications of the natural world, logically precedes understanding use. Use made sense relative to culture, not the other way around. Of course, use and culture mutually reinforced each other, and each was open to modification by historical events, as were the relationships between them. But this perspective required explicit and detailed ethnography. It required more than a rudimentary facility with a local language. It necessitated a comparative framework in which knowledge systems from different cultures could be used to illuminate one another and, beyond this, to discover any shared underlying structures of significance. And it required careful historical analysis.

The boards, however, had little tolerance for, and even less willingness to engage in, extended cultural or historical analyses. Focused on immediate difficulties of fish and game management, they were uninterested in systems of meaning. Driven primarily by the problem of sustained yield, they were not adequately apprised of the ecological role humans played in the structure and functioning of any ecosystem. They exported their lack of ethnographic enthusiasm to the subsistence research division of the Department of Fish and Game, which responded with surveys of numbers of animals killed and estimates of amounts necessary for subsistence. Culture remained an uninvited guest.

Sharing

Criterion 7 comes closest to recognizing of the importance of culture. Sharing among hunter-gatherers, as Ingold argues, "constitutes the common purpose that people *bring into* the productive process itself." Sharing is not the end of production; it is the reason for production. Exploitation and use of wild plants and animals are secondary to cultural appropriations of nature. People hunt and fish not simply for calories but to meet their social obligations to others. The majority of these obligations are based on kinship. Since the material and the cultural cannot be teased apart, "pattern of use" cannot be an unqualified, objective measure of subsistence. This point, widely supported in the ethnography of arctic and subarctic peoples, would have required the boards of fisheries and game to rethink all eight customary and traditional use criteria. It would have also required a different anthropological commitment. Once in regulations, however, and once the machinery of government took on the task of defining other peoples' customs, traditions, and associated uses, the possibility of such rethinking receded.[82]

Native peoples in Alaska and elsewhere in northern North America also hunted and fished to meet their social obligations to the animals themselves. This aspect of subsistence hunting was completely overlooked in the boards' eight-factor test. Sharing was not simply between humans. It was a larger practice that included animals. Animals shared themselves with humans; they gave themselves to hunters, as Mr. Nageak and Mr. Leavitt both said of whales. As non-human persons, then, animals have their own reasons for giving themselves to humans, but in return humans must observe a variety of ritual practices to satisfy the debt. A Yup'ik hunter noted:

> The animals, birds and plants have an awareness, and we treat them with the same respect we have for ourselves…When we bring animals into our houses, we treat them as guests…We thank them for having been caught and believe their spirits will return to their gods and report about how they are cared for. If the animals are treated well, then those gods will provide more of the same.[83]

For completeness, I should mention the eighth criterion that provides a concatenation of elements—economic, cultural, social, nutritional—associated with patterns of resource use. None of these are clearly separable in native thought, and in their current form simply replicate Western divisions—native customs and traditions notwithstanding.

"Seeing Like a State"

The first point of the preceding anti-customary-and-traditional use worksheet is this: the boards' criteria should have been provisional, with new historical and ethnographic research augmenting and perhaps superseding them, a recommendation Lonner made early in the process.[84] Even a cursory glance at the available literature would have called the customary and traditional use criteria into question. Nevertheless, with this set of criteria as a checklist, the boards of fisheries and game would decide the numbers of animals they would allow rural, mostly native, people to hunt, kill, and eat, "in amounts sufficient to provide for…customary and traditional uses." The boards explicitly set an inviolable threshold for such uses, which was not founded on any customary and traditional practices. "In no instance," the boards insisted, will subsistence harvests "jeopardize or interfere with the maintenance, on a sustained yield basis, of a specific fish stock or game population."[85] This meant that customary and traditional uses that violated, or appeared to violate, the sustained yield principle would be automatically excluded from discussion—especially uses that exhibited a long-term, customary pattern of overharvest, even if such uses were efficient, near a user's residence, based on knowledge passed between generations, and characterized by sharing. It also meant that the boards themselves had to understand and manage for species-specific sustained yield—a biologically and ecologically questionable practice, but required by the state's constitution.[86]

The relationships between native customary and traditional use and sustained yield were never fully articulated. In some areas of Alaska, "overharvesting" may have been "traditional." Steller's sea cows were dramatically reduced in numbers by aboriginal hunting along the northern Pacific Rim and then driven to extinction by European fur traders in the 1760s. Aboriginal Aleuts significantly reduced the numbers of sea otters. Iñupiat eliminated musk ox and Dall sheep from certain regions of Alaska. They then turned to other species. One traditional pattern of use appears to have been overexploitation of one species with a subsequent shift to less desirable species. Mass slaughter techniques were also traditionally used, such as driving caribou into lakes where they were more easily speared. The boards wished to eliminate these traditional practices even as they sought to preserve other, more tractable practices. Conforming to and confirming principles of sustained yield came before custom and tradition.[87]

In other areas of Alaska, "overharvesting" was the result of European technology coupled to traditional practices. Various caribou herds on the Seward Peninsula, and populations of sheep in the western Brooks Range, began to decline in the late-nineteenth century, as native peoples adopted firearms and moved into new areas in search of food and hides for clothing. Burch notes a "perfect correlation between the spread of firearms, on the one hand, and the demise of the caribou and sheep populations, on the other." He argues that new hunting technology allowed for the first time an opportunity to overharvest these animals on a regional scale. With firearms, Iñupiat could "employ their basically highly rational approach to hunting with a weapon that was technologically superior to anything they had ever possessed."[88] The result was regional overkill.

Unwilling to understand the cultural and historical complexity of hunting and gathering, the boards of fisheries and game resorted to a thorough simplification of use to facilitate decision-making and control. Their ultimate "administrative utopia," to use James Scott's phrase, was one of a well-ordered and well-regulated hunt in which every animal, as well as every hunter, was accounted for. Customary and traditional use provided one template for such order. It allowed facts to be aggregated into *classes* of phenomena—patterns of use, residential patterns, patterns of sharing, seasonal patterns, number of animals killed, and so on.[89] Such classes of phenomena were of interest to the state, although they did not match up with native concepts or the practices that resulted therefrom. Natives and non-natives, with the aid of state-employed anthropologists, were reconceptualized as "users" of "resources" and made equivalent with each other. The eight-factor C&T test facilitated the equivalency at the expense of recognizing cultural difference and historical change. The problem was that cultural worlds remained outside of the state's narrowed field of vision. The uniqueness of those worlds, however, was not simply lost from view; it was never within view. The state, through its officials and with the imprimatur of anthropologists, redefined subsistence to mean what it wanted it to mean.

Anthropologists in Alaska learned to see like a state, to borrow Scott's title to his fine book on high modernism. For Scott, high modernism is characterized by the practice of taking the complexities of the world and simplifying them in order to make natural and social phenomena legible to bureaucrats. Bureaucracies have their own standards of legibility and their own methods of rendering the world intelligible. In Alaska, those standards and methods were predominantly employed for purposes of resource control. Resource control, however, required control of relatively unknown native people, whose lives and practices must become legible in order for the bureaucracy to operate effectively. If native peoples were obscure to the state, they were less open to manipulation and control. How did the State of Alaska come to understand its people and their use of natural resources? This is where anthropologists were implicated.

Aspects of social life that could be quantified—harvest numbers, meals per week of subsistence-caught foods, distance from a village to a hunting camp, community size—came to stand for other aspects of social life. The result was a radical loss of cultural complexity. Isolating certain variables in order to domesticate them led to an uncritical reliance on those variables as reliable markers of resource use. By restricting its focus to things that could be counted, applied anthropology in Alaska ran the risk of anthropological irrelevance, at the same time that it uncritically supported the administrative goals of the state. If the goal was to understand Yup'ik resource use, then the risk of irrelevance was high. If the goal was simply to count the numbers of seals or geese Yup'ik hunters kill in any given year, then the relevance was for the state, not for local peoples, and anthropological understanding was lost along the way.

The second point of the anti-customary-and-traditional use worksheet is that customary and traditional use determinations were prescriptive rather than descriptive. They were used for particular administrative ends, even as the method of gathering cultural facts was flawed and their presentation misleading. The boards' observations of local cultural knowledge, and the statistical grid which organized those observations, were primary. The ultimate if unstated aim was to *produce* cultural systems that could be organized to assist the administration of natural systems. Despite the involvement of social scientists, adequate comprehension of cultural systems was never of genuine concern. Patterns of use would supersede culture, simply because such patterns eased the burden of administration. The anthropological arrangement of customary and traditional uses into utilitarian patterns facilitated state control.

Anthropologists were thus complicit in the state's desire to encourage natives to adhere to presumed patterns of use. They presented no viable alternatives to the boards of fisheries and game for consideration. They ignored the larger literature of their discipline and engaged in neither productive debate nor reworking of old ideas. Rather than developing new and independent analyses of custom and tradition, they remained constrained by their own bureaucratically mediated categories of use. In this way, agency anthropologists found their efforts limited, as they themselves participated in limiting the behaviors of others.

The third point is that, if Alaska were to become an efficient open-country fish and meat market based on sustained yield, then the boards of fisheries and game had to find a means to rationalize allocations of fish and game. Native custom and tradition fit the bill. Or at least with a little symbolic gerrymandering could be made to fit the bill.

The Awkward Play of Knowledge Systems

Recent research on hunter-gatherers, of the sort not typically sponsored by governments, has shown two surprising facts. Surprising, in any case, to anthropologists and world-systems scholars who formerly thought hunting and gathering would simply disappear, distorted by encompassing states or absorbed by the encroaching world system. The first fact is that hunter-gatherers exist into the twenty-first century. The second is that they have learned to manipulate state and capitalist systems for their own benefit. Their own agency has not disappeared, even as their cultures have been mistakenly presumed as either lost or transformed into something unrecognizable. Resilience and adaptability appear as central features of contemporary hunter-gatherers, which result in a kind of cultural change perhaps largely, though not exclusively, orchestrated by local meanings. As Marshall Sahlins remarks, we must "examine how indigenous peoples struggle to integrate their experience of the world system in something that is logically and ontologically more inclusive: their own system of the world."[90]

In the context of government-sponsored anthropology in Alaska, one would never know there were indigenous systems of the world. What one would know is that snow machines, airplanes, high-powered rifles, boats, global positioning systems, and other such devices have been incorporated into subsistence hunting and fishing practices, but not how they were incorporated, that is, how they entered into local symbolic systems, or with what meanings, or with what cultural effects. Of interest to the state is whether these devices result in increased hunting pressure on this or that population of animals, and the anthropology followed right along, counting, tallying harvests, and developing statistics of use. Instrumentality, efficiency, and function, not symbolism, meaning, and culture, have become the focus of anthropological effort—indigenous ontology be damned, which is to say, reduced to a utilitarian value in the Western cultural order.

The result was a failure to understand on at least two fronts with a variety of real-world consequences. By ignoring cultural variability across space and time, anthropologists constructed interpretations for government decision makers that blurred cultural differences and failed to account for historical change. Decision makers for both state and federal governments thus made resource decisions in Alaska based in part on faulty anthropological constructions. By failing to modify definitions of "traditional" and "customary" practices, government-employed anthropologists appeared stuck in theoretical modes of thought that could not accommodate the dynamics of cultural persistence *or* change.

In addition, by limiting ethnographic discussions to truncated descriptions of natural resource use, anthropologists passed over the opportunity to explain how "use" is conditioned by, indeed makes sense relative to, culture. They also avoided confronting the assumptions embedded in the phrase "natural resources." As part of the non-human world, natural resources may have seemed self-evident to resource managers, but to those with non-Western notions that posit human-like qualities to animals and to natural phenomena such as wind, a very different set of assumptions is involved. In regulatory contexts, anthropologists led decision makers to believe that cultural worlds—in this case, native worlds—were simpler and more static than they actually were, reinforcing the notion that historical change is Western and that cultural stasis—or, at best, a kind of lurching acculturation—falls to everyone else. A similar position in the natural sciences—one that asserted, for example, older ideas of ecosystem equilibrium and orderly natural succession and ignored contemporary ideas of ecosystem dynamics—would not be tolerated. But for applied anthropology in Alaska, outdated ideas were taken as settled, with no need to revisit or revise them.

The division of subsistence increasingly referenced, over the decades since its inception, not ideas debated in the international anthropological literature, but ideas generated within the work of the division itself. Subsistence research in the Department of Fish and Game became a closed system. Agency anthropologists built and then continued on a particular path, and increasingly cited themselves as the foundation for that path. With few exceptions, they ignored other trends in anthropology and relevant historical approaches to subsistence issues, and instead generated customary and traditional worksheets and an admittedly impressive array of statistical data on animals killed in Alaska. The eight-factor test has proven long-lasting, outlasting, in fact, the foundations of its rationale.

An absence of a wider scholarly community of careful critics, however, meant that conceptual improvements to applied anthropology in Alaska were mostly lacking. Critical commentary did not come from the bureaucracies in which applied anthropology was embedded. The structures of power and meaning within a bureaucracy are other than those associated with scholarly critique and may not recognize how such critique inevitably improved research. Trying to crack open a bureaucratic construct of reality to provide input is a daunting task, made more daunting because the powers-that-be could simply brush aside any critical commentary. Researchers themselves became part of the power structure and participated, for their own purposes, in the economics-based and policy-driven generation of knowledge. Such appears to be the fate of applied subsistence research in Alaska.

Perhaps the cautionary tale from this relatively obscure corner of bureaucratically driven anthropological research is quite simple. If, as Sahlins says (citing

Boas), the seeing eye is the organ of tradition, then the corollary for anthropology, including applied anthropology, must be this: The first resort is to see out of the corner of your eye.

Notes

1 *North Slope Alaska Federal Subsistence Regional Advisory Council Meeting, Public Meeting, Volume II, Barrow, Alaska* (March 8, 2011), 115. Nageak refers to Karen Brewster, ed., *Whales, They Give Themselves: Conversations with Harry Brower, Sr.* (Fairbanks, AK: University of Alaska Press, 2004). Transcripts of Regional Advisory Council meetings are available at http://alaska.fws.gov/asm/racdetail. Transcripts from all ten Councils have been underutilized in anthropological and historical assessments of resource use in Alaska.

2 *North Slope Alaska Federal Subsistence Regional Advisory Council Meeting, Public Meeting, Volume II, Barrow, Alaska* (March 8, 2011), 116–117.

3 State regulations on customary and traditional use determinations can be found at 5 Alaska Administrative Code 99.010. The federal regulations can be found at 50 CFR 100.16.

4 On the failure of Western economic models to adequately account for hunter-gatherer economies, see John Gowdy, ed., *Limited Wants, Unlimited Means: A Reader on Hunter-Gatherer Economics and the Environment* (Washington, DC: Island Press, 1998). The classic statement on the issue is Marshall Sahlins, *Culture and Practical Reason* (Chicago, IL: University of Chicago Press, 1976).

5 I have heard the napkin story from several sources, including from some of those who were in attendance. See the Alaska Joint Boards of Fisheries and Game Policy on Subsistence Resolution 81-9-JB, available at www.adfg.alaska.gov.

6 See for example James A. Fall, et al., "Cook Inlet Customary and Traditional Subsistence Fisheries Assessment" Federal Subsistence Fishery Monitoring Program, Final Project Report No. FIS03-045 U.S. Fish and Wildlife Service, Office of Subsistence Management, Fishery Information Services Division, Anchorage, Alaska (2004). Although this report provides some historical material, it fails to engage the central problem of how to conceptualize custom and tradition and their transformations. See also the following Alaska Department of Fish and Game white papers, available at www.adfg.alaska. gov: "Customary and Traditional Use Worksheet: Dall Sheep in GMU 19, McGrath Area"; "Customary and Traditional Use Worksheet: Chisana Caribou Herd, GMU 12, Upper Tanana-White River Area"; "Customary and Traditional Use Worksheets, Upper Copper and Upper Susitna River Area, Nonsalmon Finfish Species and Prince William Sound Salmon"; "Customary and Traditional Use Worksheet, King Crab and Tanner Crab, Prince William Sound Management Area: And Other Background Information." One searches in vain in such publications for anything approaching an anthropological assessment, that is, an assessment that takes seriously local cultural determinations of use. The state subsistence division publications on customary and traditional use from the early 1980s on characteristically avoid any discussion of culturally defined uses. For example, Sverre Pedersen, Terry L. Haines, and Robert J. Wolfe, "Historic and Current Use of Musk Ox by North Slope Residents, with Specific Reference to Kaktovik, Alaska," Technical Paper No. 206 (Alaska Department of Fish and Game, 1991) has 18 pages of text only three sentences of which are devoted to culture. These kinds of publications form the basis for state and federal decision makers' determinations of resource use.

7 On bureaucracies and anthropology, see Julie Cruikshank, *Do Glaciers Listen?: Local Knowledge, Colonial Encounters, and Social Imagination* (Vancouver: UBC Press, 2005), Matthew S. Hull, "Documents and Bureaucracy" *Annual Review of Anthropology* 41 (2012), and *Government of Paper: The Materiality of Bureaucracy in Urban Pakistan* (Berkeley, CA: University of California Press, 2012), David Jenkins, "Object Lessons and Ethnographic Displays: Museum Exhibitions and the Making of American Anthropology," *Comparative Studies in Society and History* 36:242–270 (1994), Josiah McC. Heyman, "Putting Power in the Anthropology of Bureaucracy," *Current Anthropology* 36:261–287 (1995), and "The Anthropology of Power-Wielding Bureaucracies," *Human Organization* 63:487–500 (2004), David Mosse, *Cultivating Development: An Ethnography of Aid Policy and Practice* (London: Pluto Press, 2005), Paul Nadasdy, *Hunters and Bureaucrats: Power, Knowledge, and Aboriginal-State Relations in the Southwest Yukon* (Vancouver: UBC Press, 2003).

8 See "The Right to Subsist: Federal Protection of Subsistence in Alaska." Alaska Federation of Natives Report on Subsistence, 2010. www.nativefederation.org/publications/subsistence. The quotes are from a letter and attachments to Secretary of the Interior, Kenneth Salazar, dated January 21, 2010, found in that report.

9 For a discussion of complex social interactions between Native Americans and Europeans resulting in a common world, even as that world included mutual misapprehensions, see Richard White, *The Middle Ground: Indians, Empires, and Republics in the Great Lakes Region, 1650-1815* (Cambridge: Cambridge University Press, 1991).

10 See James A. Fall, "The Division of Subsistence of the Alaska Department of Fish and Game: An Overview of Its Research Program and Findings," *Arctic Anthropology* 27: 68–92 (1990). Fall provides a history of subsistence research conducted by the Alaska Department of Fish and Game, including work organized by the customary and traditional eight-factor test. He does not discuss the origin or theoretical grounding of the eight factors. Robert J. Wolfe, "Local Traditions and Subsistence: A Synopsis from Twenty-Five Years of Research by the State of Alaska," Technical Paper No. 284 (Alaska Department of Fish and Game, 2004) offers an overview of twenty-five years of state-sponsored anthropology and concludes that culture is relatively unimportant in understanding subsistence traditions. Polly Wheeler and Thomas Thornton, "Subsistence Research in Alaska: A Thirty Year Retrospective," *Alaska Journal of Anthropology* 3:69–103 (2005) provide a more balanced assessment, but erroneously locate the basis of the eight factors in Division of Subsistence research. Thomas Thornton, "Subsistence in Northern Communities: Lessons from Alaska," *The Northern Review* 23:82–102 (2001) takes state and federal agencies to task for their "myopic focus on fish and wildlife harvest allocation issues." Frank Norris, *Alaska Subsistence: A National Park Service Management History* (Anchorage, AK: U.S. Department of the Interior, 2002) details the legislative history related to customary and traditional uses. Karen J. Atkinson, "The Alaska National Interest Lands Conservation Act: Striking the Balance in Favor of 'Customary and Traditional' Subsistence Uses," *Natural Resources Journal* 27:421–440 (1987) takes the position that Alaska has failed to provide a subsistence priority, as required under ANILCA, because it ignores cultural issues. Jeremy David Sacks, "Culture, Cash or Calories: Interpreting Alaska Native Subsistence Rights," *Alaska Law Review* 12:247–291 (1995) promotes the rather shrill notion that basing subsistence on culture rather than need is "dangerous." Jack B. McGee, "Subsistence Hunting and Fishing in Alaska: Does ANILCA's Rural Subsistence Priority Really Conflict with the Alaska Constitution?" *Alaska Law Review* 27:221–255 (2010) offers an idiosyncratic legal interpretation that takes customary and traditional uses as open to legal definition, ignoring cultural issues.

11 Thomas Lonner, personal communication, January 25, 2012. "Racist" was Lonner's term to describe the boards of fisheries and game. Fall, "The Division of Subsistence," and Wheeler and Thornton, "Subsistence Research in Alaska," locate the origin of the criteria with the board of fisheries, and not with Lonner, thereby privileging institutional arrangements at the expense of a more complicated and accurate history. See Lonner, "The Spider and the Fly: American Dominion and the Survival of Alaska Native Subsistence," presented to The Alaska Native Review Commission (October 1984), in which he acknowledges his authorship of the criteria.

12 See Robert J. Wolfe, et al., "Subsistence-Based Economies in Coastal Communities of Southwest Alaska," Technical Paper No. 89 (Alaska Department of Fish and Game, 1984).

13 David Mosse, "Anti-Social Anthropology? Objectivity, Objection, and the Ethnography of Public Policy and Professional Communities," *Journal of the Royal Anthropological Institute* 12:935–956 (2006), 938, 941.

14 Alaska Statehood Act, July 7, 1958, Public Law No 85-508, 72 Stat. 339, as amended; Presidential Proclamation, January 1959, 72 Stat. 339. This act mirrored Section 8 of Alaska's District Organic Act of 1884, which provided

> [t]hat the Indians or other persons in said district shall not be disturbed in the possession of any lands actually in their use or occupation or now claimed by them but the terms under which such persons may acquire title to such lands is reserved for future legislation by Congress.
>
> *Act of May 17, 1884, ch. 53, § 8, 23 Stat. 24, 26*

15 U.S. Bureau of Census, "U.S. Census of Population: 1960. Vol. 1, Characteristics of the Population. Part 3, Alaska" (Washington, DC: U.S. Government Printing Office, 1963).

16 See generally David S. Case and David A. Voluck, *Alaska Natives and American Laws*, second ed. (Fairbanks, AK: University of Alaska Press, 2002), 156 ff. The nuclear explosions were conceived as part of the Atomic Energy Commission's Plowshares Program and called Project Chariot. Edward Teller personally pitched the project in Alaska; see Dan O'Neill, *The Firecracker Boys: H-Bombs, Inupiat Eskimos, and the Roots of the Environment Movement* (New York, NY: St. Martin's Press, 1994). Radioactive waste left at the site near Point Hope in the late-1950s, apparently part of an experiment tracing the spread of radioactivity, was discovered in the early 1990s; for an account of Iñupiat responses to the discovery see Joslyn Cassady, "A Tundra of Sickness: The Uneasy Relationship between Toxic Waste, TEK, and Cultural Survival," *Arctic Anthropology* 44:8–97 (2007). On Alaska native history generally, see Donald Craig Mitchell, *Sold American: The Story of Alaska Natives and Their Land, 1867-1959* (Hanover, NH: University Press of New England, 1977). For a discussion of the pipeline, see Peter A. Coates, *The Trans-Alaska Pipeline Controversy: Technology, Conservation, and the Frontier* (Fairbanks, AK: University of Alaska Press, 1993 [original, Bethlehem, PA: Lehigh University Press, 1991]).

17 Public Land Order No. 4582, 34 Fed. Reg. 1025 (1969).

18 Case and Voluck, *Alaska Natives and American Laws*, 157.

19 As Case and Voluck point out (*ibid.* p. 165, note 64), Alaska natives perceive the 1971 payment to be in compensation for extinguishing their claim to 365 million acres, which comes to about $3/acre; the United States contributed $462.5 million and the State of Alaska contributed $500 million. For a rough historical comparison, the Neutral Tract of the Cherokee Nation in Kansas sold for $1/acre in 1871; see Richard White, *Railroaded: The Transcontinentals and the Making of Modern America* (New York, NY: W.W. Norton, 2011), 60. The issue of shifts in property concepts, from communal to corporate

ownership, in effect substituting alien for native forms, is barely discussed in Alaska anthropology. Some of the social and cultural consequences of such a shift, however, are explored in Joseph Jorgenson, *Oil Age Eskimos* (Berkeley, CA: University of Calf Press, 1990); Robert D. Arnold, *Alaska Native Land Claims* (Anchorage, AK: The Alaska Native Foundation, 1976); Donald Craig Mitchell, *Take My Land, Take My Life: The Story of Congress's Historic Settlement of Alaska Native land Claims, 1960–1971* (Fairbanks, AK: University of Alaska Press, 2001); and Steven Langdon, "From Communal Property to Limited Entry: Historical Ironies in the Management of Southeast Alaska Salmon," in *A Sea of Small Boats*, John Cordill, ed. (Cambridge, MA: Cultural Survival, 1989).

20 "Subsistence Committee Recommendations Concerning Personal Use Fishery." Alaska Board of Fisheries Resolution #81-93-FB (1981).

21 See Cathy Baker, Amy Arvideson, and Darryll R. Johnson, "Defining 'Customary and Traditional': Perspectives from Legislative History and Case Law," (unpublished manuscript, College of Forest Resources, University of Washington, 1991).

22 Congressional Record—House, November 12, 1980, H29278.

23 For one prominent issue involving mismanagement of the Western Arctic Caribou Herd, see Department of Interior. *Native Livelihood and Dependence: A Study of Land Use Values through Time* (Special Report prepared by North Slope Borough Contract Staff for the National Petroleum Reserve in Alaska, 1979), 18; Theodore Catton, *Inhabited Wilderness: Indians, Eskimos, and National Parks in Alaska* (Albuquerque, NM: University of New Mexico Press, 1997), 207–208. As with their American counterparts, the Canadian Wildlife Service blamed Inuit and Dene for the decline of caribou populations; see John Sandlos, *Hunters at the Margin: Native People and Wildlife Conservation in the Northwest Territories* (Vancouver: UBC Press, 2007).

24 5 Alaska Administrative Code 99.010.

25 Case and Voluck, *Alaska Natives and American Laws*, 286.

26 Thomas Lonner, "Perceptions of Subsistence and Public Policy Formation in Alaska," Paper presented to Society for Applied Anthropology, Edinburgh, Scotland (April 1981), 3–4.

27 "Policy Statement on Subsistence Utilization of Fish and Game," Alaska Board of Fish and Game and Commissioner, Alaska Department of Fish and Game, #76-3-JB (1976).

28 "Policy Statement on the Subsistence Utilization of Fish and Game," Alaska Boards of Fish and Game, Policy #79-5-JB (1979).

29 "Subsistence: A Position Paper," Alaska Department of Fish and Game (November 24, 1980), 6, citing the 1976 edition of *Webster's Seventh New Collegiate Dictionary*. The *Oxford English Dictionary* may have been a better choice. Note that although the oppositions were arranged along a continuum, the discussion did not specify how an opposition—kinship vs. individual, say—could transform along it.

30 Sahlins, *Culture and Practical Reason*, 7. Sahlins is speaking of kinship, but the point is clearly general and apposite in this context.

31 Lonner, "Perceptions of Subsistence and Public Policy Formation in Alaska."

32 "Subsistence: A Position Paper," 4–6.

33 "Subsistence: A Position Paper," 7.

34 John D. Speth and Katherine A. Spielmann, "Energy Source, Protein Metabolism, and Hunter-Gatherer Subsistence Strategies," *Journal of Anthropological Archaeology* 2:1–31 (1983). For a detailed discussion of nutritional value of caribou, see also Lewis R. Binford, *Nunamiut Ethnoarchaeology* (New York, NY: Academic Press, 1969). The Subsistence Division attempted to gather some data on local diets. See Daniel C. Thomas, "The Role of Local Fish and Wildlife Resources in the Community of Shaktoolik,

Alaska," Technical Paper Number 13, Alaska Department of Fish and Game, Division of Subsistence (January 15, 1982). Thomas describes a "diet calendar" which was distributed to ten Shaktoolik households and was intended to document the relative use of locally harvested vs. store bought foods. Compare, however, Grace M. Egeland, Lori A. Feyk, and John P. Middaugh, "Use of Traditional Foods in a Healthy Diet in Alaska: Risks in Perspective," *State of Alaska Epidemiology Bulletin* 2(1) (January 1998).

35 "Selected Policies of the Board of Fisheries: Joint Board's Subsistence Policy" (1982). 5 AAC 99.010.

36 Such distrust has continued into the twenty-first century. See for example the petition from Kootznoowoo, Inc., the native corporation for the village of Angoon, Alaska (2010; supplemented 2011, available at http://alaska.fws.gov/asm/issue.cfml). Kootznoowoo argued that the board of fisheries provided 15 sockeye salmon per household per year for Angoon residences, an amount the board apparently invented as an amount necessary for subsistence, with no data to support that number. The board simultaneously allowed a substantial bycatch of sockeye in the commercial fishery of over a 100,000 fish. Kootznoowoo petitioned the Secretary of Agriculture to exert extraterritorial jurisdiction, stop the bycatch, and require the state and federal governments to honor the subsistence priority. The petition was deferred to allow a local solution, with a three-year deadline.

37 Marshall Sahlins, "The Sadness of Sweetness," in *Culture in Practice: Selected Essays* (New York, NY: Zone Books, 2000), 540.

38 And remained like that. A 2012 Division of Subsistence publication, "Subsistence in Alaska: A Year 2010 Update," notes the nutritional value of wild foods:

> The annual rural harvest of 316 pounds per person contains 185% of the protein requirements of the rural population (that is, it contains about 94 grams of protein per person per day; about 51 grams is the mean daily requirement). The subsistence harvest contains 31% of the caloric requirements of the rural population (that is, it contains about 698 Kcal daily, assuming a 2,250 Kcal/day mean daily requirement).

Roy J. Shepard, however, reported that Canadian Eskimo require 3,600 kcal/per twenty-four-hour period during summer and winter caribou hunts. *Human physiological work capacity*. International Biological Programme Synthesis Series 15 (Cambridge: Cambridge University Press, 1978).

39 On notions of animal sentience among northern peoples, see, e.g., Waldemar Bogoras, *The Chukchee*. The Jesup North Pacific Expedition. Memoir of the American Museum of Natural History Vol. 7 (Leiden: E.J. Brill, 1904–1909), Robert Brightman, *Grateful Prey: Rock Cree Human-Animal Relationships* (Berkeley, CA: University of California Press, 1993), Marc Brightman, Vanessa Grotti, and Olga Ulturgasheva, eds., *Animism in Rainforest and Tundra: Personhood, Animals, Plants and Things in Contemporary Amazonia and Siberia* (Oxford: Berghahn Books, 2012), Irving Hallowell, "Ojibwa Ontology, Behavior, and Worldview," in *Culture in History: Essays in Honor of Paul Radin*, Stanley Diamond, ed. (New York, NY: Columbia University Press, 1960), Paul Nadasdy, "The Gift of the Animal: The Ontology of Hunting and Human-Animal Sociality," *American Ethnologist* 34:25–43 (2007), Frank Speck, *Naskapi: Savage Hunters of the Labrador Peninsula* (Norman, OK: University of Oklahoma Press, 1935), Adrian Tanner, *Bringing Home Animals: Religious Ideology and Mode of Production of the Mistassini Cree Hunters* (New York, NY: St. Martin's Press, 1979), and Piers Vitebsky *The Reindeer People: Living with Animals and Spirits in Siberia* (Great Britain: Harper Collins, 2005).

40 Ann Fienup-Riordan, "*Yaqulget Qaillun Pilartat* (What the Birds Do): Yup'ik Eskimo Understanding of Geese and Those Who Study Them," *Arctic* 52:1–22 (March 1999), 5.

41 Cruikshank, *Do Glaciers Listen?*

42 Huntington, et al., "The Significance of Context in Community-Based Research: Understanding Discussion about Wildfire in Huslia, Alaska," *Ecology and Society* 11 (2006).

43 Natcher, et al., "Notions of Time and Sentience: Methodological Considerations for Arctic Climate Change Research," *Arctic Anthropology* 44(2):113–126 (2007).

44 Cruikshank, *Do Glaciers Listen?*, 256.

45 Timothy Ingold, "Hunting and Gathering as Ways of Perceiving the Environment," in *Redefining Nature: Ecology, Culture, and Domestication*, Roy Ellen and Katsuyoshi Fukui, eds. (Oxford: Berg, 1996), 130–131.

46 Harvey A. Feit, "The Ethno-Ecology of the Waswanipi Cree—Or How Hunters Can Manage Their Resources," in *Cultural Ecology: Readings on the Canadian Indians and Eskimos*, Bruce Cox, ed. (Toronto: McClelland and Stewart, 1973), cited in Ingold, "Hunting and Gathering as Ways of Perceiving the Environment," 131.

47 Ingold, "Hunting and Gathering as Ways of Perceiving the Environment," 132–133.

48 Tom Lonner, *Memorandum to Subsistence Section, Alaska Department of Fish and Game* (April 28, 1980).

49 Alaska Statute 16.05.258 requires the boards to identify fish stocks and game populations "that are customarily and traditionally taken or used for subsistence" and to determine "whether a portion of a fish stock or game population…can be harvested consistent with sustained yield." If so, "the board shall determine the amount of the harvestable portion that is reasonably necessary for subsistence uses…." See www.adfg.alaska.gov/sf/publi cations/index, for a listing of C&T descriptions and ANS worksheets. Speaking in the code of acronyms and initialisms has become characteristic of government bureaucracies. Perhaps the code speeds things along by avoiding cumbersome phrases. But it also has other effects: the code produces mini black boxes whose meaningful contents remain unexamined. Acronyms and initialisms pretend a consensus of meaning which may dis appear if the boxes were opened for viewing.

50 Thomas Lonner, "What Did Subsistence Ever Do to Make Government So Mad?!," (October 3, 1982), 4–5. Paper prepared at the request of *Alaska Native News*.

51 Thomas D. Lonner, "A.D.F.&G.'s New Section," *Fish Tales & Game Trails*, Alaska Department of Fish and Game (May/June 1979), 9.

52 5 Alaska Administrative Code 01.597 (1980).

53 "Joint Boards of Fisheries and Game Policy on Subsistence," Resolution 81-9-JB (1981).

54 The Federal Subsistence Board, granted authority to act on behalf of the Secretary of the Interior and Secretary of Agriculture, comprised three rural members and the regional directors of the following agencies in Alaska: The U.S. Fish and Wildlife Service, National Park Service, Bureau of Land Management, U.S. Forest Service, and the Bureau of Indian Affairs. The Board took over control of subsistence on Federal lands after the state failed to meet provisions under ANILCA.

55 Criterion 1 was later refined to recognize shifting migratory patterns of various animals as outside a user's control. Discussions of profound historical alterations to native subsist ence practices appeared in subsistence division research, e.g., Chapter 4 of Wolfe et al., "Subsistence-Based Economies in Coastal Communities of Southwest Alaska."

56 Robert A. McKennan, "Athapaskan Groups of Central Alaska at the Time of White Contact," *Ethnohistory* 16:335–343 (1969), 336. The reference is to Richard Slobodin, "Band Organization of the Peel River Kutchin," *Bulletin of the National Museum of*

Canada no. 179, *Anthropological Series* no. 55 (Ottawa: Queen's Printer and Controller of Stationery, 1962).

57 James W. VanStone, "The Yukon River Ingalik: Subsistence, the Fur Trade, and a Changing Resource Base," *Ethnohistory* 23:199–212 (1976), 201–202.

58 Shepard Krech III, "The Eastern Kutchin and the Fur Trade, 1800-1860," *Ethnohistory* 23:213–235 (1976), 219.

59 Richard K. Nelson, *Hunters of the Northern Forest: Designs for Survival among the Alaska Kutchin* (Chicago, IL: Chicago University Press, 1973), 147–148.

60 Ibid., 148–149.

61 Dorothy Jean Ray, *The Eskimos of Bering Strait, 1650-1898* (Seattle, WA: University of Washington Press, 1975), 98.

62 For a history of the spread of tobacco, see Marcy Norton, *Sacred Gifts, Profane Pleasures: A History of Tobacco and Chocolate in the Atlantic World* (Ithaca, NY: Cornell University Press, 2008). On the fur trade, see P.A. Tikhmenev, *A History of the Russian-American Company*, Richard A. Pierce and Alton S. Donnelly, trans. and eds. (Seattle, WA: University of Washington Press, 1978 [1861]), and John R. Bockstoce, *Furs and Frontiers in the Far North: The Contest among Native and Foreign Nations for the Bering Strait Fur Trade* (New Haven, CT: Yale University Press, 2009).

63 Bockstoce, *Furs and Frontiers in the Far North*, 350–353, citing E.P Bertholf, "Report of Second Lieut. E.P. Bertholf, R.C.S." in *Report of the Cruise of the U.S. Revenue Cutter Bear and the Overland Expedition for the Relief of Whalers in the Arctic Ocean from November 27, 1879 to September 13, 1898* (Washington, DC: Government Printing Office, 1899), 103–104.

64 Theodore Catton, *Inhabited Wilderness*, 170 ff. See also *Native Livelihood and Dependence*, 121 ff.

65 Kaj Birket-Smith and Frederica de Laguna, *The Eyak Indians of the Copper River Delta, Alaska* (København: Levin & Munksgaard, 1938). Reprint edition (New York, NY: AMS Press, 1976).

66 On resource variation, predictability, and subsistence flexibility, see Douglas B. Anderson, Wanni W. Anderson, Ray Bane, Richard K. Nelson, and Nita Sheldon Towarak, *Kuuvaŋmiut Subsistence: Traditional Eskimo Life in the Latter Twentieth Century* (Kotzebue, AK: U.S. National Park Service, 1998 [1976]). Robin Ridington makes a similar argument, "Tools in the Mind: Northern Athapaskan Ecology, Religion, and Technology," in *Circumpolar Religion and Ecology: An Anthropology of the North*, Takashi Irimoto and Takako Yamada, eds. (Tokyo: University of Tokyo Press, 1994).

67 Kathryn Taylor Morse, *The Nature of Gold: An Environmental History of the Klondike Gold Rush* (Seattle, WA: University of Washington Press, 2003), 70.

68 National Research Council, *Wolves, Bears, and Their Prey in Alaska: Biological and Social Challenges to Wildlife Management* (Washington, DC: National Academy Press, 1997), 28. See Catton, *Inhabited Wilderness*, and Morse, *The Nature of Gold*.

69 David C. Natcher, Orville Huntington, Henry Huntington, F. Stuart Chapin III, Sarah Fleisher Trainor, and La'Ona DeWilde, "Notions of Time and Sentience: Methodological Considerations for Arctic Climate Change Research," *Arctic Anthropology* 44:113–126 (2007), 117. See also Bernard Saladin d'Anglure's discussion of seasonal variations, "Brother Moon, Sister Sun, and the Direction of the World: From Arctic Cosmography to Inuit Cosmology," in *Circumpolar Religion and Ecology: An Anthropology of the North*, Takashi Irimoto and Takako Yamada, eds. (Tokyo: University of Tokyo Press, 1994). Notions of time and kinship are germane to seasonality but too large a topic to include here. See, however, Robin Fox, *The Tribal Imagination: Civilization and the Savage Mind* (Cambridge: Harvard University Press, 2011) for a discussion.

70 See, e.g., Dorothy Jean Ray, "Nineteenth Century Settlement and Subsistence Patterns in Bering Strait," in *Ethnohistory of the Arctic: The Bering Strait Eskimo* (Kingston: The Limestone Press, 1983).

71 *Lieutenant Zagoskin's Travels in Russian America, 1842-1844: The first Ethnographic and Geographic Investigations in the Yukon and Kuskokwim Valleys of Alaska*, Henry N. Michael, ed. (Toronto: University of Toronto Press, 1967), 180.

72 Birket-Smith and de Laguna, *The Eyak Indians of the Copper River Delta*, 102. Cornelius Osgood reported that for the Tanaina, wolves were not eaten "because they were once men." Similarly, "Caribou were also once men, but the fact that they are eaten is explained by the statement that wolves have been men much more recently," *The Ethnography of the Tanaina*. Yale University Publications in Anthropology, no. 16 (Human Relations Area Files Press, 1976; reprint of 1937 edition), 37.

73 Ernest S. Burch, Jr. "Rationality and Resource Use among Hunters," in *Circumpolar Religion and Ecology: An Anthropology of the North*, Takashi Irimoto and Takako Yamada, eds. (Tokyo: University of Tokyo Press, 1994), 170.

74 Ibid., 171.

75 Ann Fienup-Riordan, *When Our Bad Season Comes: A Cultural Account of Subsistence Harvesting and Harvest Disruption on the Yukon Delta* (Anchorage, AK: Alaska Anthropological Association, 1986), 51, 188–189.

76 Wolfe, et al., "Subsistence-Based Economies in Coastal Communities of Southwest Alaska," Technical Paper No. 89, 368–369.

77 U.S. Bureau of Fisheries, *Alaska Fishery and Fur-Seal Industry Report* (Washington, DC: Government Printing Office, 1920). See also David B. Anderson and Cheryl L. Scott, "An Update on the Use of Subsistence-Caught Fish to Feed Sled Dogs in the Yukon River Drainage, Alaska," Final Report 08-250 (U.S. Fish and Wildlife Service, Office of Subsistence Management, Fisheries Resource Monitoring Program, 2010).

78 See, e.g., Robert Whallon, "Social Networks and Information: Non-'Utilitarian' Mobility among Hunter-Gatherers," *Journal of Anthropological Archaeology* 25:259–270 (2006); H. Martin Wobst, "The Archaeo-Ethnology of Hunter-Gatherers or the Tyranny of the Ethnographic Record in Archaeology," *American Antiquity* 43:303–309 (1978). For a linguistic/kinship analysis of regional networks of hunter-gatherers, see Per Hage, Bojka Milicic, Mauricio Mixco, and Michael J.P. Nichols, "The Proto-Numic Kinship System," *Journal of Anthropological Research* 60:359–377 (2004).

79 Draft Integrated Activity Plan/Environmental Impact Statement, Vol. 1, p. 370, Bureau of Land Management, Department of Interior, 2012, available at www.blm.gov/ak. Apparently, Anaktuvuk residents also chartered a plane in 1969 in search of caribou; Binford, *Nunamiut Ethnoarchaeology*, 51.

80 Richard K. Nelson, *Make Prayers to the Raven: A Koyukon View of the Northern Forest* (Chicago, IL: University of Chicago Press, 1983).

81 Claude Lévi-Strauss, *The Savage Mind* (Chicago, IL: University of Chicago Press, 1966), 9. In the mid-1970s, anthropologists specified the species used by Kuuvaŋmiut Eskimos in the Kobuk Valley in northwestern Alaska. Based on extensive and sophisticated knowledge of local environments, these people knew and used, for various purposes, twenty-one species of land mammals, five species of marine mammals, forty-five species of birds, 14 species of fish, and forty species of plants. See Anderson, et al., *Kuuvaŋmiut Subsistence*.

82 Timothy Ingold, "Notes on the Foraging Mode of Production," in *Hunter-Gatherers. II Property, Power, and Ideology*, T. Ingold, D. Riches, and J. Woodburn, eds. (Oxford: Berg, 1988). See Ingold, *The Appropriation of Nature: Essays on Human Ecology and Social Relations* (Iowa City, IA: University of Iowa Press, 1987), and Adrian Tanner, *Bringing*

Home Animals: Religious Ideology and Mode of Production of the Mistassini Cree Hunters (St. Johns: Memorial University, 1979). See also Fred R. Myers for other applicable references, "Critical Trends in the Study of Hunter-Gatherers," *Annual Review of Anthropology* 17:261–282 (1988).

83 Chase Hensel, *Telling Our Selves: Ethnicity and Discourse in Southwestern Alaska* (New York, NY: Oxford University Press, 1996), 71.

84 Lonner, "Subsistence as an Economic System in Alaska." Fienup-Riordan makes the same general point about Yup'ik subsistence practices: "[T]he more we recognize their diversity and flexibility, the more tenuous our generalizations become." *When Our Bad Season Comes*, 216.

85 "Joint Boards of Fisheries and Game Policy on Subsistence," Resolution 81-9-JB, 1981. For a discussion of the contradictory mandates of state sustained yield and the National Park Service policy of natural regulation, see Julie Lerman and Sanford P. Rabinowitch, "Preemption of State Wildlife Law in Alaska: Where, When, and Why," *Alaska Law Review* 24:145–171 (2007).

86 For a spectacular example of failed management based on sustained yield, see Dean Bavington's analysis of the Newfoundland cod fishery collapse, *Managed Annihilation: An Unnatural History of the Newfoundland Cod Collapse* (Vancouver: UBC Press, 2011). In September 2012, the U.S. Commerce Secretary declared a commercial fishery disaster for Chinook salmon in Alaska. It appears that, for this species of fish, sustained yield is a management failure, perhaps also leading to annihilation. Dan Joling, "Commerce Secretary Declares Alaska Salmon Disaster," Associated Press (September 13, 2012). For a study of failed fishery management in the United States, see David Jenkins, "Atlantic Salmon, Endangered Species, and the Failure of Environmental Policy," *Comparative Studies in Society and History* 45:843–872 (2003).

87 On Steller's sea cows see James A. Estes, David O. Duggins, and Galen B. Rathbun, "The Ecology of Extinctions in Kelp Forest Communities," *Conservation Biology* 3(3):252–264 (1989). On sea otters, see Charles A. Simenstad, James A. Estes, and Karl W. Kenyon, "Aleuts, Sea Otters, and Alternate Stable-State Communities," *Science* 200:403–411 (April 28, 1978). Jackson et al., "Historical Overfishing and the Recent Collapse of Coastal Ecosystems," *Science* 293:629–638 (2001) provide a deep history of the long-term ecological effects of aboriginal overfishing. For overviews of native productive practices, see generally J. Campbell, "Aboriginal Human Overkill of Game Populations: Examples from Interior Alaska," in *Archaeological Essays in Honor of Irving B. Rouse*, R.C. Dunnell and E.S. Hall, eds. (The Hague: Mouton Publishers, 1978), Charles E. Kay and Randy T. Simmons, eds., *Wilderness and Political Ecology: Aboriginal Influences and the Original State of Nature* (Salt Lake City, UT: University of Utah Press, 2002), Shepard Krech III, *The Ecological Indian: Myth and History* (New York, NY: W.W. Norton, 1999), Igor Krupnik, *Arctic Adaptations* (Hanover, NH: University of New England Press, 1993), Michael E. Harkin and David Rich Lewis, eds., *Native Americans and the Environment: Perspectives on the Ecological Indian* (Lincoln, NE: University of Nebraska Press, 2007), especially Ernest S. Burch's chapter, "Rationality and Resource Use among Hunters" (revised from the 1994 chapter, in *Circumpolar Religion and Ecology*), Charles Redman, *Human Impact on Ancient Environments* (Tucson, AZ: University of Arizona Press, 2000), and William Cronon and Richard White, "Ecological Change and Indian-White Relations," in *Indian-White Relations*, Wilcomb Washburn, ed. Volume IV of *Handbook of North American Indians*, William Sturtevant, ed. (Washington, DC: Smithsonian Institution, 1989). Robert Brightman, in *Grateful Prey*, provides a careful review of Cree overharvest, and Richard

White, in *The Roots of Dependency*, describes Choctaw overharvest; both were based on native engagement in the European market economy.

88 Burch, "Rationality and Resource Use among Hunters," 174.

89 Scott, *Seeing Like a State*, 80.

90 Sahlins, "Cosmologies of Capitalism: The Trans-Pacific Sector of 'The World System'," in *Culture in Practice*, 417.

PART II

The Nature of Bureaucracy

6

TRADITIONAL BUREAUCRATIC KNOWLEDGE

The Order of Rules

Governmental regulations that require the manipulation of natural systems are simultaneously regulations that manipulate human systems. Some of the most influential laws in the United States improved environments. In essence, these laws have told humans to stop certain practices. Many of those practices were associated with extractive industries, mining, timber production, industrial fishing, and also with the pollution resulting from the use of fossil fuels in power plants and automobiles. As humans altered their social and economic systems in response to these laws, water became cleaner and air became more breathable.

Such improvements, born from the environmental movement of the 1960s, became institutionalized by the efforts of bureaucrats charged with implementing laws such as the Clean Air Act (1963), the National Environmental Policy Act (1970), the Clean Water Act (1972), the Endangered Species Act (1973), and the Comprehensive Environmental Response, Compensation and Liability Act (1980), among others. Much has been written about the substantial human and environmental benefits flowing from these and similar acts. Much will also be written about the Trump administration's all too successful attempts to undo them.[1]

Let me start close to the ground, rather than in Congress and not with this or that administration. I've addressed those issues in prior chapters. I'll start here with a conference meeting in Fairbanks, Alaska, and with the claim that traditional bureaucratic knowledge emerges in the actions of bureaucrats.

Such actions are conditioned by laws, by regulations in service of those laws, and by directions provided by political appointees.[2] I'm especially interested in how bureaucrats themselves function in presumably rational bureaucratic systems, but which in the moment are often something other than rational. And I'm equally interested in local responses to the bureaucratic implementation of environmental laws and regulations, responses that range from enthusiastic acceptance to indifference to hostile rejection.

DOI: 10.4324/9781003297444-9

The myth of bureaucratic rationality (in part perpetuated by Max Weber a century ago) and the opposite myth of the incompetent bureaucrat are exposed in the day-to-day operations of bureaucrats, whose efforts are like the efforts of all humans, inconsistent, often confused, frequently successful, sometimes fruitless, and governed by forces they only dimly apperceive. Bureaucracies are rational and irrational. Bureaucrats themselves range from dismally incompetent to highly skilled. A small portion of them are extraordinary, working in fluid, sometimes chaotic, and frequently contradictory contexts that require a willingness to make decisions when outcomes are unclear and information is mixed, but within which decisions are required nonetheless.[3]

The ground where I start is a hotel conference room in Fairbanks, Alaska, in 2010, just after I joined the U.S. Fish and Wildlife Service as an anthropologist.[4] In February, it is cold, 20 below, but still relatively mild for Fairbanks. The sky is clear and snow graces the landscape. At this time of year, Fairbanks enjoys a little more than four hours of daylight. There are roughly 60 people assembled for a meeting focused on fish and wildlife regulations. Some of these people come from remote areas of Alaska: Coldfoot, Eagle, Huslia, Allakaket, and elsewhere. Others come from Anchorage or live in Fairbanks. In attendance are wildlife biologists and anthropologists who work for the State of Alaska, for the U.S. Fish and Wildlife Service, and for native corporations. Also in attendance are other federal employees who work for the Bureau of Land Management, the U.S. Forest Service, the National Park Service, and the Bureau of Indian Affairs. It is an impressive, diverse group.

I watch this meeting with some amazement as federal and state bureaucrats struggle to combine their diversified scientific understanding of natural and social worlds with the understandings of those from different cultural backgrounds, who continue to live on vast tracks of remote federal lands. The meeting style, for reasons that are still obscure to me, is based on *Robert's Rules of Order*. These rules of conduct, written by an American army officer in the 1870s, were derived from English Parliamentary procedures. My first experience of this meeting style is one of bafflement. I listen to a group of 20 or so individuals, all of whom live in remote areas of the state, and many of whom speak English as a second language. They sit at a U-shaped table, each with a microphone in front of them. These are members of two out of the ten Federal Subsistence Regional Advisory Councils. Their task is to advise the Federal Subsistence Board on matters of local hunting and fishing in Alaska. In its way, this is a strong, perhaps unique form of participatory democracy. Local people actually have a direct say in regulations that affect them. With only a few exceptions, the Federal Subsistence Board must defer to the wisdom of local advisory councils (often to the dismay of the State of Alaska, whose wildlife regulations are designed to benefit sport hunting and commercial activities).[5]

At work behind the current scene is a vast federal bureaucracy in various often obscure guises. All of the federal employees gathered in Fairbanks have endured background checks and fingerprinting, with my own messy fingerprinting conducted at a police station back in the Lower 48 prior to my gig in Alaska. Nowadays, such background fingerprinting is digitized and thus normalized, a different form of messiness, all that normalcy. All the members of a Subsistence Regional Advisory Council similarly endured background checks. As members of a federal advisory committee, their work is governed by the 1972 Federal Advisory Committee Act, and they have been vetted for suitability by low-level bureaucrats working for the White House. Such bureaucrats, of course, know nothing about life in bush Alaska, and less about subsistence. But they make the federal advisory committee choice of who is in, and who is out. How is it possible to make such decisions without any understanding of Alaskan ecologies or cultures? Simple. White House bureaucrats make such in-or-out decisions based on one of the main features of bureaucratic power: paperwork. Real people in real environments are secondary to the paper representations of them. I watch this gathering and realize how much I need to learn.

Federal and state staff, and a few members of the "public," sit in rows of chairs facing the two councils. I stand at the back of the room with another newcomer, a wildlife biologist from New Zealand, who was recently hired by the State of Alaska. He is as puzzled as I am. We watch as federal and state staff present biological and cultural analyses of regulatory changes and discuss scientific findings. "For the record," they say, "my name is…" "I work for…" and go on with their presentations that are mostly couched in acronym-laced bureaucratese.

Biological and cultural presentations typically last 10 or 15 minutes usually accompanied with PowerPoint slides showing graphs and charts and photographs. Local knowledge of environments and regulations is assumed. For an outsider, the language is opaque. The moose population in Unit 21E is growing, a federal biologist reports. Bull/cow and calf/cow ratios indicate the winter hunt season could be extended to a month. State biologists question some of the data, especially if those data call for limits to sport hunting. The talks move to predator control, green harvest tickets, resident-only hunting, controlled use areas, harvestable surplus, the state board of game, advisory councils, tribal opinions…. "There's quite a few moose out there on the landscape," one state biologist tells us, "that we could be harvesting that we're not."[6]

Later in the day, another biologist notes that a reduction of wolves by 60 to 80 percent is warranted in Unit 21E, in order to *increase* moose on the landscape. No one comments on the discrepancy between the two reports. There are plenty of excess moose, we are told, but let's kill wolves so we have even more moose.

The equation is, apparently, simple. Techniques to kill wolves range from trapping to issuing "aerial shooting permits" (permits being another means of bureaucratic control). One moose per square mile is the goal, which may well be a human-caused moose density, not a natural-system moose density. To get there, having no more than 29 to 31 wolves in the area ensures the moose population is high enough for human predation. Council members, keenly interested in preserving their communities' access to subsistence-caught animals, and similarly keen on preserving the populations of those animals, ask probing questions of the state and federal staff.

At a break, I ask the newly hired biologist to join me for coffee, and we chat for a few minutes. The coffee, dispensed from a large urn, is weak and slightly salty.

I'm interested in what brought this biologist to Alaska. So apparently is he, and he details his quite considerable research experience and wonders if working for a state bureaucracy in this part of the world is his cup of tea. He mentions that the state-oriented view of wildlife, which is to assess population sizes in order to maintain a sustainable harvest, is the least interesting part of wildlife management. "Count and kill," he says with a shake of his head. He is much more interested in northern ecologies and their intricate relationships and is concerned that he won't be able to pursue his research interests. It happens that within a month, he will quit and return to New Zealand.

After the break, state and federal anthropologists analyze what they call "customary and traditional uses" of killed animals. I've become increasingly puzzled as I've listened to descriptions of native life that are not about customs and traditions. They are about how wild game is killed and how food is prepared and eaten. It is a discussion about calories and how many pounds of wild game are *minimally* necessary for people to subsist on, from an outsider's perspective. An "eight factor" test of customs and traditions appears to be the metric of metabolic necessity, but it is not explained in any detail at this meeting. From what I gather, the approach reflects a certain functionalist anthropology from the 1970s, and I am intrigued and a little dismayed to hear it in Fairbanks in 2010, as if it captures the current state of anthropological analysis. A bureaucratic *test* for customs? Based on eight *factors*? I listen closely from the back of the room and wonder if biologists in Alaska have similarly stopped their thinking with theoretical orientations from the 1970s. If so, I've landed in a very strange bureaucratic landscape.

Anthropologist Claude Lévi-Strauss once critiqued functionalist anthropology as an analytical technique that transforms culture into a vast metaphor for digestion. I'm watching just such a transformation in real life, in which native customs become Western calories—as magical as turning water into wine. Bureaucrats are the magicians. Does anyone else notice this oddity? My boss, the federally employed anthropologist who hired me, joins me at the back of the room and I mention this critique. What seems to matter is how much one eats, not cultural relationships with animals and environments and humans. What seems to matter is what can be easily quantified. Where, I ask her with apparent alarm, is the cultural analysis? She gives me a wan smile and says something about learning the federal system. The bureaucratic world in which I have landed is odd indeed. I wonder, not for the last time, if my boss questions her decision to bring me on board. I also wonder what other sorts of magical bureaucratic transformations I'll encounter.

I write in my notebook, "Who are these people? What in the world are they talking about?" Although we share a common language, I find it difficult to follow the discussion. Later in the proceedings, other people representing commercial interests, environmental groups, or tribes approach the council to speak. Some offer PowerPoint presentations, others offer anecdotes about life in the bush. Unlike scientists, environmental advocates, or bureaucrats, native people characteristically introduce themselves by saying where they are from and to whom they are related. They then often provide a personal story of the central importance of subsistence-caught foods in their families' lives, emphasizing place and kin and continuity. They also emphasize the great respect they hold for the animals they kill and consume.

Throughout the conference, council members and federal and state staff effortlessly engage in ritualized talk: "through the chair," "the question's been called," "do I have a second," "all those opposed by the same sign." When things occasionally go awry, the proceedings are halted and *Robert's Rules* consulted to ensure that the council adheres to the "proper" order of action and engages in "appropriate" ritualized speech. Now and then, the process becomes too tangled to sort out, a veritable Babel of languages and parliamentary officiousness. The entire event is recorded and later transcribed, according to requirements of the Federal Advisory Committee Act. All presentations are available online, or rather only the English spoken portions. Discussions in native languages are elided, which is unfortunate because some of those discussions, as I later learned, turned out to include snide and often humorous commentary on the process. The ten subsistence advisory councils meet twice a year, and frequently in remote locations, where bureaucrats and local representatives attempt to communicate about shared concerns.

Research on contemporary native cultures in Alaska emphasizes how subsistence hunting and gathering persists into the twenty-first century. Much of the literature focuses on modern equipment which allowed people to continue their productive practices within the context of local, primarily kin-oriented culture. With a few exceptions, considerably less research effort focuses on how local peoples make use of government regulations to do the same thing. Yet, arguably, government regulations are more significant than modern firearms, snowmobiles, ATVs, and GPS devices, in both modifying and allowing for the cultural continuity of hunting and gathering practices.

The problem is basically this. Local peoples are forced to adapt to what they see as an incoherent system of governance and to find creative ways to both keep at bay and make use of regulations without overly damaging their own sense of who they are or derailing their own projects. Bureaucracies are famously implacable, and large state and federal bureaucracies governing wilderness in Alaska are especially so. The bureaucracies and the regulations they promulgate, however, are as much a fact of life for native peoples as are shifting migrations of caribou, growing populations of brown bears, or moose in Unit 21E.

One telling and celebrated example of such incoherence occurred in Barrow, Alaska. It predates the 1980 establishment of subsistence resource advisory councils and served to ignite native resistance to imposed regulations curtailing traditional hunting practices.

In the 1950s, the U.S. Fish and Wildlife Service determined that the Migratory Bird Treaty Act precluded Iñupiat from hunting certain species of ducks. Iñupiat hunters mostly ignored this governmental imposition. They continued to hunt ducks, according to their own cultural practices. In 1961, Fish and Wildlife Service wardens arrested two Iñupiat men for possessing ducks they had killed.

"On the morning after the arrests," writes historian Theodore Catton, "138 Barrow Inupiats appeared outside the wardens' hotel, each one with a dead duck in hand, demanding his or her own arrest as well."[7] This coordinated display of civil disobedience, along with growing discontent about imposed regulations, led to the Alaska Native Claims Settlement Act, which President Nixon signed into law in 1971, and to the Alaska National Interest Lands Conservation Act, which President Carter signed into law in 1980. These laws provided the bureaucratic basis for ecological and human control. Yet the relationships between governance and local cultures remained to be defined bureaucratically and played out culturally. I became part of that process.

Years later, *Robert's Rules of Order*, not duck-in-hand civil disobedience, appeared to be one somewhat awkward means to mediate between very different worlds. To me, this is where traditional bureaucratic knowledge resides—and can be illuminated—within that awkward space of mediation.

This is my second claim about traditional bureaucratic knowledge. If it emerges in the actions of bureaucrats, it also uneasily resides between different knowledge systems. The action-oriented aspect of bureaucratic knowledge receives most of the critical attention. The uneasy relationships between different knowledge systems receive much less critical attention. Like most systems of knowledge and action, there is nothing fixed about traditional bureaucratic knowledge. It is fluid and dynamic, even as the bureaucratic structures themselves limit the fluidity and dynamism. Various chapters try to sort out these two claims by reference to specific examples, from trees to water to fish to humans, and back again to humans.

I also invoke "traditional bureaucratic knowledge" to make a point about "traditional ecological knowledge," the subject of a later chapter. The central problem is with the notion of "tradition," and the often-uncritical acceptance of native (or more broadly, non-Western) "traditions" as if they themselves were somehow timeless. History pervades all cultures, however, even cultures that appear outside of it. Those of us who support the importance of traditional ecological knowledge, and I happen to be one of them, in the larger understanding of humanity, need to highlight how changeable, flexible, and situational such knowledge is, and to avoid making traditional ecological knowledge into some sort of fixed set of practices and beliefs.

In some ways, traditional ecological knowledge is more open to novel circumstances than traditional bureaucratic knowledge, with fewer constraints. The latter could learn from the former.

Traditional bureaucratic knowledge takes a variety of forms, as demonstrated throughout this book, depending upon the bureaucracy in question and upon how bureaucrats imagine environments and interact with the people whose lives they influence. Despite the fluidity and dynamism of bureaucratic thought, two rather static bureaucratic elements are key: *order* and *rules* to maintain that order. One element is imagined, the other imposed. Each is open to critique. Each is culturally determined. I eventually came to see *Robert's Rules of Order* as the metaphor for bureaucratic knowledge in general: putatively rational, culturally arbitrary, difficult to parse, avidly followed and just as avidly resisted, even as few people can say precisely why such rules are in place.

Over several years working in the U.S. Fish and Wildlife Service's office of subsistence management, I became a bureaucrat, if a poorly socialized one. But I was also, and perhaps primarily, an anthropologist and I found the bureaucratic culture in which I worked as exotic as Yup'ik, Iñupiat, and Athabascan cultures. Indeed, bureaucracy as a culture was something more exotic and more inexplicable than native cultures, with which I had some natural affinity and among whom I had previously worked. Of course, I use the term "exotic" not to marginalize this or that culture, nor to turn anyone into an "other," but rather to emphasize that all cultures are equally in need of explication when viewing the entire pageant of humanity, as anthropologists tend to do. We are, all of us, equally exotic, just as we are, all of us, equally banal.

I found it more congenial to imagine a world of sentient nonhuman persons with wills and motivations of their own, than to imagine, let along work within, a world of acronym-wielding bureaucrats who insisted that rules and regulations were more important than reality and acted as if the bureaucracy itself was more significant than its ostensible mission.

In Alaska, I marveled at grizzly bears who comprehended human speech and reacted to human discord. I admired bowhead whales who selflessly volunteered to die, according to the Iñupiat, to feed humans. And I stood upon glaciers which expanded and contracted in response to human moral failings. At least I endeavored to understand these cultural orientations because I endeavored to understand their meaningful and symbolic forms.[8] The cultural orientations of government bureaucrats, by contrast, I struggled against, sometimes openly so, even as I eventually mastered the federal system and learned to speak in terms of calf/cow ratios and the eight factor test of customary and traditional use.

Still, in the command-and-control culture of a federal land management bureaucracy, I was frequently at odds with the knotted control of flows of information, in which people accrued power by secreting information to themselves and to their increasingly small groups of intimates. Sharing information on a "need to know" basis may work in a military context, but in an organization dedicated to preserving natural systems a more rational course is open, transparent communication. I preferred to let information fly every which way, to our state counterparts, to native corporations, to other federal agencies, and to the public. I preferred a sort of radical truthfulness.

My preference, however, was a minority preference, voiced from below. In the bureaucratic caste system, I could easily be ignored; anthropologists are frequently ignored in such contexts. I continued to question the bureaucratic hierarchy, in which micromanaging "supervisors" arbitrarily demanded compliance of "their staff," and who thought it unnecessary to provide any rationale for an assigned task or to engage in any productive debate. And I was at odds with the profoundly anti-intellectual tenor of much of the work, which allowed no critical commentary intended to improve or streamline existing practices and discouraged research independent of imagined natural orders and imposed bureaucratic rules. Bureaucracies have limited fields of vision, and those with unconstrained vision are often outcast by those who administer rules and impose order.

Bureaucracies, I came to realize, promote complacency and compliance. They often function by stifling creativity. They weed out those who question the status quo and elevate those who preserve it. Bureaucracies are self-perpetuating social forms, as Max Weber recognized a century ago, which tend to persist even as their reason for being is lost. New ideas, proffered descriptions of social organization or structures of significance, impromptu cultural analysis, or simple enthusiasm for intellection are somehow intimidating in a bureaucracy—at least they were for some of the top managers in the office I happened to work in. Governmental bureaucracies can be awkward places to work for those of us who are neither complacent nor compliant.

I have since developed a more nuanced view of bureaucracies and bureaucrats, but without abandoning a critical or creative stance—my own sort of self-preservation within the system. I now see that traditional bureaucratic knowledge is more permeable and less monolithic than I had experienced in Alaska. But it is constrained and often contradictory. Hierarchical social forms limit knowledge production. Those in power can simply tell those without power to be quiet. At the other extreme, bureaucratic incentives for awards and promotions often serve to limit knowledge production. It is hard and individually perilous to question prevailing orthodoxy if one is anxious to improve one's position and increase one's pay. It is especially hard for those in the senior executive service in the federal government to go against the political grain. They can simply be moved to a new position in retaliation for any political apostasy and without further explanation. In such circumstances, there is often a self-imposed silence on opinions that challenge those in control.[9]

Such silencing exists in many cultural situations. Religious officials rarely tolerate strongly dissenting points of view, for example. But secular bureaucracies in their preservation of order and imposition of rules have institutionalized both the control of knowledge and the control of dissent. The world may put cultural meanings at risk, as anthropologist Marshall Sahlins remarks, simply because the world is always different than how it is imagined. But at least from my experience within bureaucracies, the world must work harder to put traditional bureaucratic knowledge at risk, precisely because of institutionalized control.

All is not lost. The fact that anthropologists and other social scientists are hired to work in governmental organizations of natural resource management lends hope to the eventual understanding of the complexities of coupled human and natural systems. The pages in this volume attempt to contribute to that understanding. They explore from various angles what I've come to call "traditional bureaucratic knowledge." They describe imagined orders and imposed rules, knowledge production and knowledge constraint, and highlight environmental successes and failures. They try to understand and to knock slightly off-center traditional bureaucratic knowledge. They exist, these pages, between knowledge systems. My hope is that folks working on the ground in local environments with local communities will recognize the transformative power they in fact hold, despite what may be irrational direction from above.

As with the Iñupiat in Barrow (renamed in 2016 "Utqiaguik") I want ducks to survive, and I want Iñupiat to survive. Our planet is simply better off with such diversity. As I see it, there is no contradiction between biological diversity and cultural diversity. They may well require one another. Ideally, land management bureaucracies should evolve to ensure both sorts of survival, and bureaucrats within them should hasten that evolution, even if politicians must be dragged along.[10] Here's to the hastening.

Notes

1 See Eric Lipton, Steve Eder, John Branch, and Gabriella Demczuk, "This Is Our Reality Now," *The New York Times* (December 27, 2018). Rebecca Leber, "The EPA Is Planning to Jeopardize the Water Quality of 117 Million Americans," *Mother Jones* (December 11, 2018). Michael Biesecker and Mathew Brown, "The Trump EPA Moves to Rollback More Clean Air and Water Rules," *The Washington Post* (March 1, 2018).

2 Who may not know what they are doing. See Michael Lewis, *The Fifth Risk* (New York, NY: W.W. Norton & Company, 2018).

3 See for example Peter J. Balint, Ronald E. Steward, Anand Desai, and Lawrence C. Walters, *Wicked Environmental Problems: Managing Uncertainty and Conflict* (Washington, DC: Island Press, 2011).

4 In the Office of Subsistence Management. www.doi.gov/subsistence.

5 For an overview, see David S. Case and David A. Voluck, *Alaska Natives and American Laws*, second ed. (Fairbanks, AK: University of Alaska Press, 2002).

6 Western Interior Regional Advisory Council Meeting, Fairbanks, Alaska, February 25, 2010. Office of Subsistence Management. www.doi.gov/sites/doi.gov/files/migrated/subsistence/library/transcripts/upload/Region-6-25-Feb-10.pdf.

7 Theodore Catton, *American Indians and National Forests* (Tucson, AZ: University of Arizona Press, 2016), 112. See also Theodore Catton, *Inhabited Wilderness: Indians, Eskimos, and National Parks in Alaska* (Albuquerque, NM: University of New Mexico Press, 1997), and Donald Craig Mitchell, *Take My Land, Take My Life: The Story of Congress's Historic Settlement of Alaska Native Land Claims, 1960-1971* (Fairbanks, AK: University of Alaska Press, 2001).

8 The literature on notions of animal and geomorphic sentience among northern peoples is large and varied. See, e.g., Waldemar Bogoras, *The Chukchee*. The North Pacific Expedition. Memoir of the American Museum of Natural History vol. 7 (Leiden: E.J. Brill, 1904–1909), Robert Brightman, *Grateful Prey: Rock Cree Human-Animal Relationships* (Regina: Canadian Plains Research Center, 2002), Marc Brightman, Vanessa Grotti, and Olga Ulturgasheva, eds., *Animism in Rainforest and Tundra: Personhood, Animals, Plants and Things in Contemporary Amazonia and Siberia* (Oxford: Berghahn Books, 2012), Julie Cruikshank, *Do Glaciers Listen?: Local Knowledge, Colonial Encounters, and Social Imagination* (Vancouver: UBC Press, 2005), Irving Hallowell, "Ojibwa Ontology, Behavior, and Worldview," in *Culture in History: Essays in Honor of Paul Radin*, Stanley Diamond, ed. (New York, NY: Columbia University Press, 1960), Paul Nadasdy, "The Gift of the Animal: The Ontology of Hunting and Human-Animal Sociality," *American Ethnologist* 34:25–43 (2007), Frank Speck, *Naskapi: Savage Hunters of the Labrador*

Peninsula (Norman, OK: University of Oklahoma Press, 1935), Adrian Tanner, *Bringing Home Animals: Religious Ideology and the Mode of Production of the Mistassine Cree Hunters* (New York, NY: St. Martin's Press, 1979), and Piers Vitebsky, *The Reindeer People: Living with Animals and Spirits in Siberia* (Great Britain: Harper Collins, 2005).

9 Former Secretary of the Interior Ryan Zinke preemptively moved around 33 senior executives, ostensibly to improve efficiency, but more likely to remove those from positions of bureaucratic power who held views different from the incoming administration or who were promoted to their positions in the Obama administration. He gave them 15 days to move from one position to another, often to less suitable positions, sometimes to completely inappropriate positions. Experience and knowledge lost; political purity won. See Juliet Eilperin and Lisa Rein, "Zinke Moving Dozens of Senior Executives in Shake-Up," *The Washington Post* (June 16, 2017), and Joel Clement, "Secretary Zinke's Diversity Problem," *Union of Concerned Scientists* (April 3, 2018) https://blog.ucsusa.org/joel-clement/secretary-zinkes-diversity-problem.

10 For a timeline of the failure of politicians in climate change mitigation, see Paul Bledsoe, "Going Nowhere Fast on Climate, Year After Year," *The New York Times* (December 29, 2018).

7

BUREAUCRATIC MANAGEMENT OF WILDLIFE

Wolves in the State of Alaska

Anthropologist David Graeber, a strong critic of bureaucracies, notes that "Contrary to popular belief, bureaucracies do not create stupidity." He goes on to say that bureaucracies "are ways of managing situations that are already inherently stupid because they are, ultimately, based on the arbitrariness of force."[1] In wildlife management, such arbitrariness takes the form of legal enforcement—enforcement of what I call the "order of rules." By this I mean an imagined view of how ecosystems best function, and the rules and legal apparatus to stabilize that imagined view. The management of wolves in Alaska is a case in point.

The complicated science of wildlife management is often less than clear. Wildlife regulations, and the Western values in support of those regulations, can be even muddier. The State of Alaska, as noted in Chapter 4, has a constitutional provision that requires the state to manage animals for "sustained yield." In 1994, the legislature adopted an intensive wildlife management statute (Alaska Statute 16.05.255) which states that "high levels of harvest for human consumptive use in accordance with the sustained yield principle is the highest and best use of identified big game prey populations in most areas of the state." A perpetual supply of wild game is the goal, just as, under similar notions of sustained yield, a perpetual supply of timber and perpetual supply of fish are the goals. Killing predators—bears and wolves, in particular—provides one way to manage for sustained yield and high levels of harvest of other animals, such as moose and caribou. After many challenges, the Alaska Supreme Court agreed the "highest and best use" of big game animals is for human consumption. Predator control remains one means to ensure a steady supply of wild meat for humans to eat.[2]

The calculation is simple. Humans eat moose and caribou. So do bears and wolves. When moose or caribou are scarce, humans should kill wolves and bears to increase populations of animals preferred for human consumption. For people in remote areas of Alaska dependent upon wild game for subsistence, the calculation is

DOI: 10.4324/9781003297444-10

understandable. They are in effect competing with bears and wolves over moose and caribou. For those who kill moose and caribou for sport, however, the calculation is questionable. They too compete with wolves and bears over moose and caribou, but not primarily for subsistence. Whatever the motivation, moose populations in Alaska are rarely "natural," nor are the populations of their predators. Both predator and prey, and the dynamic between them, are all deeply intertwined with humans.

This fact is disguised by the very language of predator control and by the court's decision upholding such control. Regulations assert that the intent of predator control is not to eliminate wolves but to maintain them "as part of the natural ecosystem...." How is a natural ecosystem achieved? Not by leaving it alone. As a former governor of Alaska once said, "You can't just let nature run wild."[3]

For the State of Alaska, a "natural" ecosystem is achieved, first, by setting "management objectives" for predator population size, second, by ceasing predator control and closing hunting seasons when necessary "to ensure minimum population objectives are met," and third, by defining "specific geographic boundaries" for those areas destined for predator control.[4] Thus, the notion of a "natural" ecosystem is belied by massive human intervention in it. Nature running wild it is not.

Wildlife management, in other words, produces animals that are neither entirely wild nor entirely managed. Such animals, subject to both natural forces and human intervention, become semi-domesticated, blurring the line between nature and culture. Bureaucracies (and courts) are the means for such line-blurring. Management regimes, however, often act as if scientific research into animal biology and relevant ecological processes provides sufficient grounding for management decisions about wildlife. If we know enough about animal biology, the argument goes, and about how animals interact with their natural environment, then we are in a position to make adequate management decisions. With sufficient knowledge of nature, wildlife management becomes self-evident.

There is some force to this argument. Scientific assessment provides a necessary informational element to managing wildlife. But if managed wildlife is recognized as semi-domesticated, the intervention of humans must be factored into any adequate understanding of this or that population of animals—including the intervention of bureaucratic humans. The central management problem is not simply to understand animal biology and behavior and to act accordingly. Rather, the central problem is to understand the relationships between humans and wild animals, relationships that, in comparison to biological phenomena, remain disproportionately understudied.

The former Director of the Alaska Division of Wildlife Conservation was a strong advocate for killing predators (despite the title of the division he led). In 2009, then working as Assistant Commissioner, he published a brief article in *Sportsmen's Voice*. He argued that "the maximum sustainable yield principle requires that managers be willing to assume the risks and responsibilities of trying to maximize human benefits by intervening in natural systems over the longer term." He further asserted that managing wildlife is "deeply rooted in the culture and traditions of Alaska." The former Wildlife Conservation Director goes on to say, "Central to these traditions is the premise that mankind has not only the ability, but also the obligation to manipulate natural systems for the benefit of people, as well as for the benefit of the resource itself."[5]

His article was accompanied by an advertisement for T-shirts (Med-2XL) advocating killing wolves. Emblazoned with a wolf on a pack of cigarettes, and with a moose wearing sunglasses and holding a smoking gun, these "Limited-Edition 'Smoke-A-Pack-A-Day'" T-shirts were "the most sought after clothing item we have ever produced…a must for your next hunting trip."[6]

In this publication, the Alaska constitution, state legislation, predator control, the authority of a state office holder, court rulings, and a rather vulgar marketing expression of "deeply rooted" cultural traditions were combined in a single brief text.

Wolves, following the scientific footsteps of wildlife management, were subject to a "smoke-a-pack-a-day" mentality. Sustained yield of their prey was of paramount management concern. Killing wolves was the means to that end. More moose were then available for human predation and consumption. That much was straightforward, if of questionable ecological wisdom. Contemporary predator control, enforced by court decision, followed decades of poisoning wolves and shooting them from airplanes, which are no longer common practices.[7] Such practices however, produced moose populations that were larger than they would have been otherwise. Alaskans came to see those artificially large populations as natural, and not a product of human decisions. When wolves got in the way of "highest and best use," the solution was to kill them.

As semi-domesticates, wolves, managed for human benefit, were not immune to other human-mediated environmental changes, nor to other contradictory forms of governmental control. I will describe one such contradiction of control I found both amusing and telling. Amusing, in any case, from the human side of things; less so from the perspective of wolves. And telling because governmental bureaucrats, uninformed of each other's actions, acted in ways that baffle.

The odd environmental circumstance in Alaska was this: dog lice, probably from sled dogs, infected wolf populations. First detected among wolves on the Kenai Peninsula in the early 1980s, dog lice spread to wolves as far north as Fairbanks. Wolves infected with dog lice react the same way dogs react, by scratching at their fur, rendering their fur unmarketable. Trappers were unhappy, because lousy wolf pelts had no value. They needed healthy animals to trap and intact furs from those animals to sell. They eventually asked the State of Alaska to intervene.[8]

The odd human circumstance was, naturally enough, bureaucratic. In the bureaucratic event, the Alaska Department of Fish & Game, with federal funds, developed a delousing program, intended to identify infected wolf packs and control louse infestations. The solution was to locate wolf dens and to drop from the air "fist size chunks of moose meat" injected with ivermectin, a common drug used to kill parasites in household pets. As part of the program, state biologists bought lousy wolf hides from trappers. They also captured 34 wolves "using helicopter capture techniques," and radio collared 29 wolves to follow the movement of some 10 to 12 packs.[9]

Ivermectin (made famous with bogus claims it treats coronavirus infections in humans) kills dog lice for about two weeks. The treatment of ivermectin-bearing moose meat flung from helicopters had to be repeated to ensure lice-free wolf packs. The overall intent of the de-lousing effort was to improve the health of wolf populations, to slow the spread of dog lice among wolves, and to ensure trappers could trap marketable fur-bearing animals. The effort, however, proved expensive and not very effective.

How does this de-lousing program square with Alaska's predator control program? It appears that the State of Alaska prefers to control, that is kill, predators only if they are healthy. Healthy wolf packs, which naturally kill and eat moose and caribou, among other animals, can then be controlled in order to artificially increase moose and caribou populations. Dropping chunks of ivermectin-laced moose meat to wolves in order to improve their health so they can be killed to increase moose populations—this may appear to some to be the height of rational wildlife management. For others, the practices are irrational and indefensible. It is in such contradictory environmental effects that ordinary bureaucratic forms are illuminated.

Wildlife managers meet their legislative mandate for sustained yield of pre-ferred prey animals even if their practices are contradictory. The former Director of Wildlife Conservation for the Alaska Department of Fish & Game appeared to solve the contradiction with his assertion that mankind is obliged to manipulate ecosystems to benefit people and resources. One wonders on what evidentiary basis he made a claim for all of mankind. He clearly did not consult Yup'ik or Iñupiat people.

One also wonders whether a benefit to wolves is simply undone with their subsequent control. Of course, the element missing from the Director's assertion is whether humans have sufficient knowledge to "intervene in natural systems over the longer term." Without such knowledge, managing wildlife because we have the ability and the obligation may result in undermining the very resources we wish to preserve. Saving wolves in order to kill them pretends a lucid bureaucratic ration-ality, and then hopes for the best.[10]

Notes

1 David Graeber, *Fragments of an Anarchist Anthropology* (Chicago, IL: Prickly Paradigm Press, 2004), 73.

2 West v. State Board of Game. Nos. S-13184, S-13343. Supreme Court of Alaska (August 6, 2010).

3 Malcolm B. Roberts, ed., *The Wit and Wisdom of Wally Hickel* (Anchorage, AK: Searchers Press, 1994).

4 West v. State Board of Game. Nos. S-13184, S-13343. Supreme Court of Alaska (August 6, 2010).

5 Corey Rossi, "Managing for Abundance: Producing a Bounty for All Users to Enjoy," *Sportsmen's Voice* 29 (Summer, 2009).

6 Ibid.

7 National Research Council, *Wolves, Bears, and Their Prey in Alaska: Biological and Social Challenges in Wildlife Management* (Washington, DC: National Academy Press, 1997).

8 Ted Spraker, "Dog Lice Found on Kenai Wolves," *Alaska's Wildlife: The Magazine of the Alaska Department of Fish and Game* (November–December 1991).

9 See Howard N. Golden, Ted H. Spraker, Herman J. Griese, Randall L. Zarnke, Mark A. Masteller, Donald E. Spalinger, and Bruce M. Bartley, "Infestation of Lice among Wild Canids in Alaska," Division of Wildlife Conservation/Department of Fish & Game www.wolfsongalaska.org/chorus/node/91. See also Craig Gardner, Kimberly Beckman, Nathan J. Pamperin, and Patricia Del Vecchio, "Experimental Treatment of Dog Lice Infestation in Interior Alaska Wolf Packs," *Journal of Wildlife Management* 77(3):626–632 (April 2013).

10 In yet another contradiction of bureaucracy, Corey Rossi, appointed by half-term governor Sarah Palin to his Director's position, was later arrested and convicted of illegal bear poaching. See Craig Medred, "The Spectacular Rise of Alaska Wildlife Manager Corey Rossi," *Anchorage Daily News* (January 15, 2012).

8

ENEMY ANCESTORS

Bureaucratic imagination is historically shallow and conceptually narrow, focused on those things that can be counted and arranged in categories and that conform to its vision of the world. However, a careful examination of bureaucratic practice shows that the knowledge systems involved are mixed, with economic, scientific, political, and cultural knowledge systems as part of bureaucratic imagination. There is nothing seamless about the mix of knowledge systems in a bureaucracy, and often such systems are at odds with one another. In some ways, bureaucracies function to exclude novel ideas, but not always successfully. I've offered here as examples forestry practices, water policy, fisheries management, and bureaucratic interaction with non-Western cultures, among others. I've argued that crises are often the only way to expand the narrow field of bureaucratic awareness. Climate change is one such crisis with all sorts of cascading effects, from severe drought, wildland fire, wildlife declines, insect apocalypse, and fisheries collapse. Here I want to suggest an expansion of bureaucratic imagination in two directions, one that focuses on the past and another that goes counter to those things "easiest to monitor, count, assess, and manage," to borrow from James Scott's characterization of administrative utopia.

I've already alluded to "enemy ancestors"—those whose actions and mal-formed beliefs about the world have left us with degraded environments. A scan of the Environmental Protection Agency's Superfund website provides a sense of the scale of the problem. The Superfund resulted from the 1980 Comprehensive Environmental Response, Compensation, and Liability Act, which created a tax on the petroleum and chemical industries, whose hazardous wastes were all too often abandoned, contaminating water and soils. That tax has been used to pay for cleaning up hazardous materials when those responsible managed to skirt responsibility. Forty years after the establishment of this act, there remain more than 1,300 identified contaminated sites that require cleanup. The hazards are acute. Heavy

DOI: 10.4324/9781003297444-11

metals such as mercury, arsenic, and lead, and aromatic hydrocarbons such as benzene, are toxic to humans and other organisms. The list grows daily.[1] A similar scan of the U.S. federal government website devoted to abandoned mine lands shows an even larger problem, with more than 500,000 abandoned mines causing a variety of environmental hazards and health concerns. These include risk of exposure to arsenic, lead, and radiation.[2]

The idea of enemy ancestors needs to be planted deeply in the bureaucratic imagination. We need to recognize the failures of those who came before us. We need to find ways to avoid similar failures. We need to explicitly call out the massive hazards our ancestors left for us to deal with. To that end, I suggest developing a favorite bureaucratic tactic, namely an index: a simplified at-a-glance summary of those who left us and our world imperiled.

Such an index, the stuff of bureaucratic practice, is not easy to assemble because it responds to personal sentiments that, as Max Weber demonstrated, are inimical to a fully functioning bureaucracy. To get there also requires moving along a very un-bureaucratic path, winding through wilderness, migrations, languages, remote Aleutian Islands, tsunamis, Japanese nuclear power plants, and salmon fisheries. Crooked paths reveal more than dead-straight roads.

An unexpected connection may be found between Alaskan wilderness and Anasazi (ancient Puebloan) ruins. The connection is not as far-fetched as it first appears. Ancestors of the Athabascan language family migrated from Siberia to Alaska some 12,000 years ago, although the exact timing is open to debate. Over millennia, the original language diversified into 38 daughter languages, whose speakers spread out over a vast area of Alaska, Canada, northern California and the southwestern United States. Seven of these daughter languages are found in the southwest and include six Apache languages (Western, Mescalero, Chiricahua, Jicarilla, Lipan, and Kiowa Apache) and Navajo. Ancestors of the people who speak Apachean languages migrated from the Subarctic relatively recently, perhaps no more than 500 years ago, probably in a series of migrations.[3]

One Apachean language, Navajo, gave us the word "Anasazi," which means "enemy ancestor," or "ancient enemy," a term that has fallen from favor in recent years. But to the extent that any of our ancestors have left us with environments unsuited for humans and the organisms with which we share the planet, the word "Anasazi" fits all too well. We live with the decisions of those who came before us, and many of those decisions turn out to be ill-informed and wrongheaded. Some of our ancestors were willfully blind to the obvious. We live with the results of their blindness. They are indeed our ancient enemies.

To return to Alaska in 2010. Hanging in my office there was a wilderness poster based on a photograph of an out-of-the-way bay of Kiska Island, part of the westernmost end of the Aleutian chain far out in the Bering Sea. In the foreground of the photograph surrounded by knee-high sedge sits a rusting 75-millimeter Type 88 anti-aircraft gun, which the Japanese used to guard an army base they built at Gertrude Cove in 1942. In the shallow bay in the distance, one can make out the wreck of a Japanese freighter, the Borneo Maru, destroyed by American bombs.

Here, on this small island around this small cove, humans have left a wilderness in their historical wake. But the message of abandonment is not one of saving wilderness or celebrating Aleutian landscapes so much as remembering spirits. The caption of the poster reads "Ghosts of World War II." The site, a National Historic Landmark, is part of the World War II Valor in the Pacific National Monument.[4]

The Japanese invasion of the Aleutian Islands was meant as a distraction, intended to pull American forces away from Midway and the anticipated naval battle. The invasion required an American response that included rounding up over 800 Unangam Aleuts from nine villages on nearby islands for their own safekeeping. These people, whose houses and churches were burned to keep them out of Japanese hands, were involuntarily relocated to southeast Alaska, where they stayed for two years in dismal living conditions. Kiska Island itself had no native people at the time of the Japanese invasion. The Japanese bombed Americans at Dutch Harbor on Unalaska Island; the Americans bombed Japanese at Gertrude Cove on Kiska Island. American forces then descended upon isolated Kiska—nearly 35,000 troops, 95 ships, and 168 aircraft—but the Japanese had already departed. The battle of Midway proved decisive in the Japanese loss of the war, coupled to the devastating atomic bombing of Hiroshima and Nagasaki. Eighty years later, the ghosts remain at Kiska Island, memorialized. Archaeologists continue to find prehistoric artifacts in the debris of bomb craters, linking past and present, wilderness and world system.[5]

The Kiska Island poster reminds us that the larger world intrudes into remote wilderness regions at unexpected moments and with unexpected consequences— consequences that highlight past decisions, and just as often call those decisions into question. Perhaps this is a common historical phenomenon. Perhaps an intruding world is implicated in all governmental bureaucracies, which then becomes apparent as we rethink the actions of our ancestors. It is, in any case, a theme of this book by a federal bureaucrat. I sort through moments from which we should learn deep historical lessons, and I am hopeful that we can learn those lessons. But I remain pessimistic. I write in March 2020, during a worldwide coronavirus pandemic that has killed and will continue to kill millions of people, and I despair at the incompetence of the current U.S. president and the stuttering response from many of his handpicked cabinet secretaries who lead major federal bureaucracies. Our descendants may come to see all of them as enemy ancestors.

The examples at hand, however, are these: cancers in Downwinders years after atomic testing in Nevada, declines of Yukon River Chinook salmon despite robust data gathering and rational management practices, catastrophic wildfires the result of extreme wildfire suppression in our nation's forests, failures to provide adequate supplies of water to western cities, and the almost complete extirpation of Atlantic salmon from its natal river systems. But our attention often appears too shallow, too self-focused, too driven by immediate political machinations and the bureaucracies implementing those machinations, to make better collective decisions based on these historical lessons.

One important historical moment was the March 11, 2011, earthquake and resulting tsunami that devastated parts of northeastern Japan and caused radiation to leak from the collapsed Fukushima Daiichi nuclear power plant. The full consequences of this moment, however, will take time to play out. The power of the earthquake shifted the Earth's axis by 25 centimeters and changed its rotation, shortening the length of a day by 1.8 microseconds. Over 14,000 people were killed by the tsunami. A month after the initial event, 15,000 people were still missing. The tsunami destroyed well over 125,000 buildings. Around 4.5 million households were left without electricity. Over 200,000 people were evacuated from the area around the power plant. An unknown number will suffer from radiation exposure. Soon after the earthquake and tsunami, low levels of radiation from damaged nuclear reactors drifted to North America on prevailing winds. Off the coast of Japan, ocean currents move east and northward to the Bering Sea, and then shift southward along the coast of Alaska. For many years to come, migratory birds, marine mammals, ocean currents, and wind currents could be affected by, or carry the effects of, the meltdown at Fukushima. Thus does the environment alter humans. Thus do humans modify their environment. Remote Kiska Island is not immune.[6]

Although scientists assert that the North Pacific and Bering Sea fisheries will not be affected by Fukushima radiation, the public perception is of heightened risk. One concern is the large number of Japanese-origin hatchery fish. North Pacific hatcheries release over five billion salmon juveniles annually. The majority comes from Japan. These fish feed and mature in the Pacific Ocean and Bering Sea, where they mix with wild fish. As a mix, wild and hatchery fish are caught by industrial fishers for the global market. In 2007, an estimated 513 million salmon were caught in the North Pacific by Alaskan fishers (213 million), Russian fishers (213 million), and Japanese fishers (76 million). The overall value of the North Pacific commercial salmon fishery is substantial, with a 2007 wholesale value of $2.2 billion.[7] By 2020, the fishery was in sharp decline, with an estimated 322.5 million salmon caught, even as the 5 billion annual hatchery releases have been stable since 1993.[8]

The ocean is large. We are told it will adequately dilute radiation. Iodine-131, with a relatively brief half-life of eight days, won't pose an extended problem. Cesium-137, with a half-life of 30 years, is another matter. As a recent article in *Science* observes, an internal dose of Cesium-137 "could induce mutations, stunt growth, and cause reproductive defects" in the organisms so exposed. The ocean disperses. Animals, feeding along the food chain, accumulate. At the bottom of the food chain, phytoplankton and zooplankton accumulate radioisotopes to high levels. Humans, at the end of the chain for many fish species, may be the recipients of the accumulation. Warnings about heavy metals in the flesh of whales, seals, and walruses are common. We do not yet know about radiation from the Fukushima Daiichi nuclear power plant accumulating in marine mammals, fish or in the humans who consume them, although current projections are for minimal risk. Nor do we know how the floating debris swept out to sea with the retreating tsunami will retain contamination. The 2011 Fukushima nuclear disaster is in its way a vast experiment on the relations between nature and culture, an experiment that replicates the similar disasters at Windscale in the United Kingdom (1957), Three

Mile Island in the United States (1979), and Chernobyl in the Ukraine (1986). In 2019, 450 nuclear reactors were in operation worldwide; 58 were under construction; and 63 hand been shut down between 2005 and 2019. Some countries have since taken steps to stop facilities under construction, and to limit or outright ban future nuclear facilities.[9]

The world intrudes. Choices in one part of the planet affect other parts. Because the world is in motion, it spreads the effects of human action far beyond its local origin. Humans too are in motion, bringing along their own unpredictable relations to themselves and to the natural world. Perhaps future interactive maps of nuclear power plant meltdowns will show a radiation index for each disaster and its effects on both near and remote regions of the planet. Such maps need to be conceived not in two or three dimensions, but in four dimensions with time forming a crucial informational component. We can then chart the consequences of choices made by our ancestors. If those choices prove to be shortsighted and wrongheaded and result in further damage to all who share the planet, we will need to rethink and rename them. "Mistakes of our ancestors" seems too mild an epithet. We need an index appropriate to those mistakes. Call it the Anasazi Index.

Notes

1 www.epa.gov/superfund/superfund-cercla-overview. For an extended analysis of CERCLA and its limitations, see David Jenkins, Joanne Bauer, Scott Brunton, Diane Austin, and Thomas McGuire, "Two Faces of American Environmentalism: The Quest for Justice in Southern Louisiana and Sustainability in the Sonoran Desert," in Joanne Bauer, ed., *Forging Environmentalism: Justice, Livelihood and Contested Environments* (New York, NY: M.E. Sharp, 2006).

2 www.abandonedmines.gov/.

3 See James Kari and Ben A. Potter, eds., *The Dene-Yeniseian Connection*, Anthropological Papers of the University of Alaska, New Series, Volume 5 (Fairbanks, 2010), R.G. Matson and Martin P.R. Magne, *Athapaskan Migrations: The Archaeology of Eagle Lake, British Columbia* (Tucson, AZ: University of Arizona Press, 2007), Ripan Singh Malhi, et al., "Distribution of Y Chromosomes among Native North Americans: A Study of Athapaskan Population History," *American Journal of Physical Anthropology* 137:4 (2008).

4 www.nps.gov/articles/aps-v10-i1-c5.htm.

5 See Dean Kohlhoff, *When the Wind Was a River: Aleut Evacuation in World War II* (Seattle, WA: University of Washington Press, 1995); Rachel Mason, "You Can't Go Home Again: Processes of Displacement and Emplacement in the 'Lost Villages' of the Aleutian Islands," *Alaska Journal of Anthropology* 8:2 (2010).

6 See Richard Hindmarsh, ed., *Nuclear Disaster at Fukushima Daiichi: Social Political and Environmental Issues* (New York, NY: Routledge, 2013).

7 North Pacific Salmon Fisheries Economic Measurement Estimates Version 1.2 Prepared by the Research Group for Wild Salmon Center (December 2009). https://wildsalmoncenter.org/wp-content/uploads/2016/02/Sal%E2%A8%B5%EE%B7%AE%E2%9B%BC_Economic_Valuation.pdf.

8 "North Pacific Salmon Catches Decline in 2020 to Lowest Levels in Four Decades," North Pacific Anadromous Fish Commission. https://yearofthesalmon.org/wp-content/uploads/2021/05/6-NewsReleasesIYS.pdf.

9 Sara Reardon, "Fukushima Radiation Creates Unique Test of Marine Life's Hardiness," *Science* 332:6027 (April 15, 2011), 292. See also Teppei J. Yasunari, Andreas Stohl, Ryugo S. Hayano, John F. Burkhart, Sabine Eckhardt, and Tetsuzo Yasunari, "Cesium-137 Deposition and Contamination of Japanese Soils Due to the Fukushima Nuclear Accident," *Proceedings of the National Academy of Sciences of the United States of America* 108:49 (December 6, 2011), and Hideki Kaeriyama, "Radioactive Cesium in the North Pacific after the Fukushima Dai-ichi Nuclear Power Plant Accident," *North Pacific Anadromous Fish Commission Technical Report No 17*: 202–204 (2021).

9

TO SAVE THE SPIRITUAL

Environmentalists often make strong pleas for preserving what few natural areas are left on the planet. One wonders where such places may be, if they are defined by the absence of human presence. The 1964 Wilderness Act tried to make such a definition:

> A wilderness, in contrast with those areas where man and his works dominate the landscape, is hereby recognized as an area where the earth and its community of life are untrammeled by man, where man himself is a visitor who does not remain. An area of wilderness is further defined to mean in this Act an area of undeveloped Federal land retaining its primeval character and influence, without permanent improvements or human habitation, which is protected and managed so as to preserve its natural conditions and which (1) generally appears to have been affected primarily by the forces of nature, with the imprint of man's work substantially unnoticeable; (2) has outstanding opportunities for solitude or a primitive and unconfined type of recreation; (3) has at least five thousand acres of land or is of sufficient size as to make practicable its preservation and use in an unimpaired condition; and (4) may also contain ecological, geological, or other features of scientific, educational, scenic, or historical value.[1]

Note the four categories of bureaucratic significance "hereby recognized": (1) forces of nature and the related absence of "man," (2) solitude and unstructured play, (3) landscape manageability, and (4) a grab bag of various "values." I want here to explore several of these categories and to propose a different way of thinking about them.

I start with a poster the Southern Utah Wilderness Alliance produced in support of preserving wilderness in the southwest United States. This poster hangs in my

DOI: 10.4324/9781003297444-12

office, across from an image of remote Kiska Island in the Aleutians. It shows a transcendent desert scene from Cedar Mesa in southeastern Utah, based on one of David Muench's photographs. In large lettering at the bottom is the word WILDERNESS. Along the left margin, a quote from Wallace Stegner: "The spiritual can be saved."

What I find especially compelling about this poster of spiritual wilderness is the image. It is a picture of an ancient Puebloan ruin tucked into a cliff, serene, timeless, built from small blocks of sandstone and fashioned to fit neatly into a natural formation. It is a picture of human presence, of the domesticated wild. Humans, or at least their handiwork, figure prominently in this image of wilderness.

It is a mistake, however, to think that the message is unambiguous, promoting and connecting spiritualism and wilderness ideals. And it begs the question: whose spiritualism? Archaeologists have shown that many Puebloan communities used up their resources until they were forced to abandon their homes. Making initial use of nearby wood supplies for buildings and cooking fires, for instance, some communities necessarily ranged farther afield, harvesting less valuable wood over greater distances as nearer sources disappeared. Over time, usually hundreds of years, sometimes just a few generations, the stress on resources became too great, wood was depleted, and people moved on to more hospitable areas. Changes in climate often hastened the process. Marauding cannibalistic invaders from the south may have provided further incentive to move, as stories of their arrival terrified local inhabitants.[2]

As an icon of modern wilderness values, perhaps also expressive of modern spiritual values, the image of a Puebloan ruin deep in the North American desert fails to alert the viewer to two essential facts: humans use resources, and humans abuse other humans. From this perspective, the striking wilderness image shows the failure of a group to people to adapt to their environment, which they abandoned. The Southern Utah Wilderness Alliance poster thus comes with an implicit warning. Even those people who live their lives in close, intimate relationships with the encompassing world may leave a wilderness behind them, having used up that world and decamped for other, more favorable climes.

I often thought about ancient Puebloan ruins as I researched contemporary problems of toxic oilfield waste in southern Louisiana and sustainable development in southern Arizona—part of an international research effort mentioned in the introduction. I wondered if our modern categories, such as solitary/social, natural/unnatural, wild/domesticated, country/city, wilderness/civilization, and their cognates, would have made any sense to the inhabitants of those ruins. They made little sense to me.

A wilderness includes those "areas where man and his works" *interpenetrate* the landscape, just as a city contains the interpenetration natural world. Separating the categories revealed more about human desires than about either wilderness or city. I wondered if we moderns will be forced to leave our homes, either through exhausting our resources or from the abuse of other humans. We see both occurring almost daily on our shared planet.

My colleagues on the international project—political scientists, sociologists, and anthropologists in China, India, and Japan—studied problems of environmental disasters and resource uses in their own countries. We wanted to know how local cultural values entered the policy making process, whether environmental concerns of local people were being adequately addressed by policymakers, and whether there were any commonalities across markedly different study sites in these four countries. The short answers are that local cultural values rarely enter national environmental policy making, and that many policy commonalities we found were based on the command-and-control efforts of governments pursuing their own economic agendas. Ultimately, we wanted to give advice to policymakers and bureaucrats on how to better understand and accommodate local peoples, whose perspectives had been marginalized. I'm not sure we succeeded.

When I arrived at the University of Arizona in Tucson to begin this project, I was immediately struck by how fast the city was expanding. My first impressions were not favorable, even as I grew to love the city. A great-grandson of Mormon pioneers, I had grown up in Salt Lake City, and I watched as that city encroached upon agricultural lands until suburbs filled the valley from rim to rim. Tucson, set in a much wider basin, seemed bent on the same ill-considered course. Sprawl, the unofficial motto of western cities must be, until there is no more room to sprawl.

As I drove around in the first few days after my move to Tucson to get a sense of a city I had last seen in 1976, I could tell that Tucson had recently leaped beyond its old margins. Storage facilities, usually at the edges of towns where land is cheap, were in the middle of suburbs. Bars with names like Boondocks, once in the boon-docks, were likewise in the middle of suburbs. Motels built in the 1960s and 1970s, with gaudy signs still intact, no longer greeted travelers as they entered town; they were deep within the town, often hard to find. As Tucson grew, its heritage of beautiful and graceful Spanish architecture transformed into something else. Many expensive homes in the foothills above the city looked to me like World War II bunkers: thick, squat, menacing, and ugly, an architectural vision of doomsday, with the sun as the enemy. The interiors of such houses are light and spacious, which disguises the small gunsight windows, the large overhanging forms blocking out the sun, the yards into which no one ventures in the summer heat. The whole basin was chock-a-block with air-conditioned shopping malls.

Two years later, before leaving Tucson and eventually alighting in Alaska for a job with the U.S. Fish and Wildlife Service, I hiked into the foothills above the city one late afternoon. It had become my daily habit to look about and imagine what the basin below me would be like in 25 years, or after 100 years, or after 500 years. Looking over the city I thought of distant environments linked to this basin, via dams and canals and water pumped from the Colorado River uphill 230 miles through the desert to Tucson—a system that took 20 years to build at a cost of $4.7 billion.

I recalled the many weeks I had spent kayaking long stretches of the Green River in northern and central Utah, the Dolores River in Colorado and Utah, and the San Juan River on the southern border of Utah, all of which flow into the

Colorado River. On these trips, I floated past well-known Fremont, Basketmaker, and Puebloan ruins, past ruins known only to archaeologists, and past ruins known and loved by those who explore the mysteries of the desert for their own purposes but are rarely spoken about. These are wilderness rivers, perhaps, controlled by dam releases, traveled by tens of thousands each year on rafts, canoes, and kayaks, and administered by various bureaucracies.

Let me explain what I mean. I also want to acknowledge and subvert David Graeber's comment about bureaucracy and play: "What ultimately lies behind the appeal of bureaucracy is the fear of play."[3] This is also what Max Weber referred to as the "complete eradication of love, hate, and all purely personal sentiments from administrative tasks." Bureaucracies and their administrative tasks are the antithesis of play, even as play may be a result of those tasks.

I knew the National Park Service allowed 23,000 people each year to float the Colorado River through the Grand Canyon between June and September and recalled that 20,000 people floated every year through Desolation Canyon on the Green River. As a conservative estimate, I figured another 10,000 floated the upper Green River through the Gates of Lodore, and at least 10,000 more on the San Juan River. Humans average one to two quarts of urine per day.

As I walked up a steep ridge above Tucson, a favorite hike of mine off the usual trails, I worked out the calculations. Using the higher average (boaters, in my experience, drink a lot of beer) and figuring the boating season to be three months long for all the rivers, I calculated that 63,000 people times 90 days times two quarts of pee each day yields 2,835,000 gallons of human urine directly entering the system each year. In the jargon of river measurement, that's more than 8.7 acre-feet of pee. The actual figure is probably double or triple that. River runners have become part of each river's ecosystem, and in a distant way part of Tucson's ecosystem.

I looked down at the city and imagined nearly nine acre-feet of urine, which I thought to be relatively benign stuff, mostly water, but an interesting if quirky index of human use of the wilderness. At least such an index called into question the idea of solitude and the absence of man, even as it highlighted unconstrained recreation. I first called it the Wilderness Pee Index (WPI), pronounced "whipee."

I considered only the main rivers. I had not calculated yearly estimates for many smaller and heavily traveled rivers such as the Gunnison, Yampa, Salt, and Dolores Rivers. I thought that each wilderness river could be assigned a wilderness index, based on the amount of human urine entering the system. One could then choose to float a river based on that index and the relative human use it represented. I wondered if I could entice this or that bureaucracy, the National Park Service, perhaps, or the Bureau of Land Management, to adopt my index in its planning documents.

A few months later I realized that a better name with the same acronym is the Wilderness Pharmaceutical Index. Human urine, as it turns out, is not so benign. Recent research has shown detectable levels of a variety of contaminants in U.S. rivers, some portion of which is derived from human urine and is not screened by wastewater treatment plants. Low but persistent levels of antibiotics, antidepressants, steroids, reproductive hormones, over-the-counter-drugs, and other pharmaceuticals have been found in more than 100 U.S. streams. "Wherever humans live or visit," one research report noted, these kinds of pharmaceuticals "are constantly infused into the environment."[4]

Over time, aquatic organisms may be affected in a variety of ways. For example, selective serotonin reuptake inhibitors (SSRIs), such as Prozac and Zoloft, effective at elevating human mood, also alter the spawning behavior of bivalves. Serotonin is in fact common in vertebrates and invertebrates, and low persistent levels of SSRIs that humans excrete into rivers may have a variety of nervous system effects on animals exposed to them. Such effects may be difficult to distinguish from changes brought about by natural selection. Much more research is needed to understand how aquatic organisms respond to low levels of hormonally active chemicals and other drugs designed to produce a physiological response in humans. Humans too may be at risk. As the same report noted: "A major unaddressed issue regarding human health is the long-term effects of ingesting via potable waters very low, subtherapeutic doses of numerous pharmaceuticals multiple times a day for many decades."[5]

That day above Tucson, however, I was unaware of the pharmaceutical complications. I reached the top of the rock ridge that allowed me to quickly gain a thousand feet in elevation and continued higher, picking my way through ocotillo and fishhook cactus, watching a Peregrine falcon hunting up a meal, and marveling at the African grasses so well-adapted to the region, and so ubiquitous, that many people mistake them for native species. I moved deliberately, stepping carefully over rocks, eyes open for one of many species of rattlesnake in the area. Looking over the city, I wondered how our descendants will cope with the decisions we and our ancestors have made. What will they think of the canal system that links Tucson with the Colorado River? How long will the canal function? When will high energy costs prohibit pumping water uphill across hundreds of miles of desert? Will the population of Tucson stabilize?

Perhaps water wars will come to the West as sources are depleted. Perhaps the population growth of western cities will reverse and cities such as Tucson, Phoenix, and Las Vegas will shrink, become ghost towns, Puebloan ruins of the modern age. In that case, years hence, the dry canal system that used to pump water uphill will become an object for future wilderness posters hanging in some other anthropologist's office, recalling an earlier time when people failed to adapt to their immediate environment and left a wilderness in their historical wake. The spiritual can then be saved, again.

Notes

1 The Wilderness Act, Public Law 88–577 (16 U.S.C. 1131–1136). Section 2(c).
2 See Christy G. Turner II and Jacqueline A. Turner, *Man Corn: Cannibalism and Violence in the Prehistoric American Southwest* (Salt Lake City, UT: University of Utah Press, 1999).

3 David Graeber, *The Utopia of Rules: On Technology, Stupidity, and the Secret Joys of Bureaucracy* (Brooklyn, NY: Melville House, 2015), 93.
4 Christian G. Daughton and Thomas A. Ternes, "Pharmaceuticals and Personal Care Products in the Environment: Agents of Subtle Change?" *Environmental Health Perspectives* 107:6 (1999), 908.
5 Ibid.

10
TRADITIONAL ECOLOGICAL KNOWLEDGE

The appearance of traditional ecological knowledge (TEK) in scientific discourse illustrates how meanings are always dependent upon prior structures of significance, a topic touched on in various chapters and further explored in Chapter 11. Such structures of significance, as they relate to land management, are often bureaucratically enforced through statutes and regulations, policy directions and on-the-ground practices. They emerge as actions by bureaucrats trying to implement this or that policy. Increasingly, they are informed by neoliberal ideologies. The relationship between traditional ecological knowledge and scientific knowledge is a particularly hazy and intriguing blend of knowledge systems. As pointed out in Chapters 4 and 5, bureaucracies of land management have a hard time with this combination.

Many anthropologists are leery of the notion of "tradition." Mary Douglas asserts that "There is no such thing as 'traditional culture.' It is misleading to speak as if it were definable and recognizable." It may seem odd that anthropologists question "tradition." But Douglas goes on to say there is nothing immutable about culture, including any "traditions."

> Culture is a dynamically interactive and developing sociopsychic system. At any point in time the culture of a community is engaged in a joint production of meaning. (This is very different from the ways that culture is described in other literatures, either as based on language, or geography, or on social class, or shared history.) In reality, the connected meanings that are the basis of any given culture are multiplex, precarious, complex, and fluid. They are continually contested and always in the process of mutual accommodation. The dialogue leads to concentrations of meaning. It is the process of self-understanding, the way a community explains itself to itself.[1]

DOI: 10.4324/9781003297444-13

Things get tricky when a community explains itself to itself by recourse to its "traditions," especially when those explanations are elevated into international discourse. The World Commission on Environment and Development, for example, in its Brundtland Commission Report, notes that indigenous "communities are the repositories of vast accumulations of traditional knowledge and experience that link humanity with its ancient origins."[2] Such an assertion is founded on a view of cultural continuity that is contradicted by historical change. But it may well form the basis of a community's explanation of itself.

Traditional ecological knowledge, as it relates to scientific knowledge, seems to get caught up in cultural practices of self-understanding and in larger claims of presumed connections to our deep past. These are very different phenomena. One is of immediate cultural concern of self-definition, the other an ideological statement about indigeneity in general. Meanwhile, the contemporary social forces at play, and the dialogue between traditional knowledge and scientific knowledge, lead to concentrations of meaning, which, not surprisingly, tend to ignore the anthropological critique.

Internationally recognized work in British Columbia from the mid-1990s highlighted indigenous knowledge and its importance in building better approaches to sustainable forest management. The Scientific Panel for Sustainable Forest Practices in Clayoquot Sound sought to understand and incorporate First Nations' perspectives on resource use and environmental planning. Its mandate from the provincial government was "to make forest practices in Clayoquot not only the best in the province, but the best in the world."[3] Recognizing that western-style forest management often resulted in degraded environments, the panel reached out to First Nations for help. Together they developed a list of 12 TEK characteristics. The Scientific Panel based this list on existing literature and in consultation with local Nuu-Chah-Nulth people. From these sources, they determined that TEK has the following characteristics: it is holistic; intuitive; qualitative; transmitted intergenerationally by oral tradition; governed by a Supreme Being; moral; spiritual; based on mutual well-being, reciprocity, and cooperation; non-linear; often contextualized within the spiritual; communal; and promoting of stewardship.[4]

There are several interesting elements of such lists, which are not unique to the Clayoquot Scientific Panel. First, by emphasizing "traditional" knowledge, the Panel ignores the history of indigenous peoples and places them in a framework of mythic timelessness—as if, unlike other modern people, they have not been changed by, or are not agents of, their own history. "Traditions," however, are clearly mutable. They morph based on any number of historical and cultural factors. In this feature, they are no different than other aspects of culture.

James Scott notes, "The term 'traditional,' as in 'traditional knowledge'…is a misnomer, sending all the wrong signals." He illustrates:

> In the mid-nineteenth century, explorers in West Africa stumbled upon groups growing maize, a New World grain, as their main staple. Although it was unlikely that the West Africans had been growing maize for longer than

two generations, its cultivation was already surrounded by elaborate rituals and myths about a maize goddess or spirit who had given them the first kernels. What was striking was both the alacrity with which they had adopted maize and the speed with which they had integrated it into their traditions.[5]

Scott suggests that high modernism, with its faith in science, needed an oppositional knowledge system which it could perceive as backward and primitive in order to secure its own dominance. For high modernism, that oppositional knowledge system existed in the realm of tradition. Non-Western traditions were a human curiosity, perhaps, but were not useful for the heavy lifting of high modernism and the Progressive Era with its emphases on science and organizational efficiency. More recently, scientists appear to embrace non-Western traditions, perhaps recognizing the limitations of scientific supremacy in human affairs. But the embrace of tradition comes at a cost, the loss of those self-same traditions and their dynamism, as they become mere fixed ciphers for a certain kind of empiricism open to scientific inquiry. I'll return to this point.

The second interesting feature of the Scientific Panel's list is this: it lumps together diverse cultures, without any understanding of the environments or social worlds within which gods, moral qualities, spiritual forces, or cyclical timeframes make sense. In a footnote, to take one example, the Scientific Panel asserts local Nuu-Chah-Nulth philosophy, emphasizing the "oneness" of everything, "differs little from that of the Quakers."[6] Perhaps this is vaguely true at some high level of abstraction, but the histories and cultures of Nuu-Chah-Nulth and Quakers remain profoundly different. To ignore those differences provides little evidence of presumed commonalities. Saying it's so doesn't make it so. One would think that hardly needs emphasizing.

The Scientific Panel also ignores the often extensive environmental modifications resulting from native productive practices, preferring instead to focus on the modern issue of sustainability. Sustainability is chimerical, with rare examples across the planet, and it has become important in present times precisely because of its unlikeliness. Thus, scientists look elsewhere for inspiration, perhaps seeing in the modern world the economic engine of our eventual demise. "Tradition," then, as in Scott's argument, retains its status as "other," but as hopeful rather than backward, filled with possibility rather than dismissal. The characteristics adduced by the Scientific Panel remain the opposite of scientific knowledge, commonly thought of as objective, quantitative, and reductionistic. The Scientific Panel, in its own report, pushes against scientific knowledge and its presumed limitations and looks elsewhere for help.

The third interesting element of the Scientific Panel's list is it takes the idea of "ecosystem" and its relationship to other cultural traditions as a given. The term and idea have become such a constant in environmental assessments that rarely is the term questioned—I've never heard it questioned in over a decade working in land management agencies. Not surprisingly, the difficulties with defining "ecosystem" are couched in terms that resemble the difficulties with defining "culture," with questions of boundedness, wholeness, systemic integrity, part/whole relations, interdependence, scale, and change as central questions. Both ecosystems and cultures may be systematic in some regards; but if so, they are open systems, which makes both understanding the system, and prediction of the system, a very large challenge.

Perhaps, as anthropologist Roy Rappaport has suggested in *Ritual and Religion in the Making of Humanity*, the notion of ecosystem will eventually offer the basis for unifying diverse understandings of the natural world and human presence within it. Ecosystems provide the context for different ways of thinking of and engaging with nature-culture relationships. However, as functioning systems, ecosystems cannot be empirically demonstrated; at best they provide a general template for scientific investigation. At the same time, for scientists and non-scientists alike, Rappaport notes, the notion of ecosystem "provides a general view of the physical world in light of which people can formulate their practical and moral relationship to it." In addition, the notion "is not only an explanation of nature but a reflection upon nature and a guide for acting in nature." Since a functioning ecosystem cannot be demonstrated, "its validity is, therefore, like the validity of all conventions, including religious understandings, a function of acceptance."[7]

A fourth element of the Scientific Panel's list deserves mention. The panel ascribes traditional ecological knowledge to oral cultures. TEK, the panel notes, is transmitted between generations by oral tradition. This characteristic seems to exclude from consideration a wide diversity of non-Western cultures which developed robust writing systems. These include the Sumerians, who invented writing more than 3,000 years ago, and the Maya and the Chinese, who independently invented their own writing systems. Even the Cherokee, whose syllabary was invented by Sequoyah in the late 1810s, would not fit securely under the panel's definition. Orality as one defining characteristic of TEK unnecessarily excludes cultures that invented systems of writing.

I'm not questioning the need for a broad discussion of how best to manage forests or other natural resources. Such a broad discussion is, in my view, necessary. I often question whether management based on economics and policy-based systems of knowledge yields the best results. We should consult people with long-term experience in the environments we try to manage and attempt to understand their points of view; indeed, they should be co-managers in many cases. There is ample evidence that indigenous peoples' lands are degrading at slower rates than lands governed by Western-style models of management. How and why indigenous lands are managed differently are important questions to pursue. These are not my questions here.[8]

What I want to emphasize is a basic anthropological point. Indigenous cultural meanings are not at play as scientists try to incorporate them into their own systems of knowledge. Indigenous cultural meanings—structures of significance— remain obscure to scientists defining TEK and building lists of such knowledge. In list-building, the Scientific Panel, like many Western environmentalists, selects an exceedingly small subset of alternative meanings to fit its own agenda and serve a pragmatic goal, as is evident in its self-definition. "The Nuu-Chah-Nulth phrase *hishuk ish ts' awalk* ('everything is one') epitomizes a holistic world view, and has been adopted by the Clayoquot Scientific Panel to describe the ecosystem management approach to forest practices the Panel recommends." The hopes of the Nuu-Chah-Nulth, and of a set of scientists, are in this sense coextensive and serve to guide government policy. Holistic ecosystem management science and an "everything is one" philosophy coincide. Yet holism and oneness are cultural orientations; they are not scientifically demonstrable or even open to scientific investigation. They are, however, open to historical and cultural analysis, as are the motivations to combine them.

Interestingly, the Scientific Panel's TEK list resembles the characteristics of deep ecology, which similarly emphasizes personal experience, holism, and the intrinsic worth of the natural world. Consider this foundational assertion:

> Deep ecology is emerging as a way of developing a new balance and harmony between individuals, communities and all of Nature. It can potentially satisfy our deepest yearnings: faith and trust in our most basic intuitions; courage to take direct action; joyous confidence to dance with the sensuous harmonies discovered through spontaneous, playful intercourse with the rhythms of our bodies, the rhythms of flowing water, changes in the weather and seasons, and the overall processes of life on Earth.[9]

All apparently good things, balance and harmony, joyous dancing, playful intercourse, connected—at least in metaphor—to flowing water, seasonal change, and all life on Earth. Yet deep ecology appears to have lost the high ground.

As an orientation to public lands management, "ecosystem services" has triumphed instead. Ecosystems services first emerged in the 1970s in an attempt to bring greater attention to biodiversity. It then became a means to impose pecuniary values on the natural world, with all the attendant theorizing from economics.[10] It is now institutionally secure in U.S. land management practices; indeed, the Forest Service website on ecosystem services asks the neoliberal question, "How can we make good stewardship profitable?"[11] Scientific effort, as was clear in the Progressive Era, could be used for profitable ends, deep ecology, in the modern era, less so. But will traditional ecological knowledge similarly be bent toward profit? Will TEK be subordinate to ecosystem services? The anthropological question is, *which knowledge system filters the other*? The economic question is, *who benefits*? The political question is, *who has the power to say*?

For my part, I question the faith and trust in our basic intuitions as we dance and harmonize with Wall Street, British Petroleum, or the former U.S. Minerals Management Service, all of which did the Texas Two-Step in a look-the-other-way-and-smile government oversight of oil industry practices in the Gulf of Mexico. One result: the Deepwater Horizon oil spill, the largest marine oil spill in history. I struggle with the reality of our economic and political systems in which economic policies outdo policies of conservation and preservation. And I find myself something of a cynic about incorporating non-Western ideas into science and land management, especially when many of those non-Westerns ideas—however matched and combined from various sources—challenge the policy status quo.

The problem is, our basic yearnings are conditioned in part by our energy consumption, made possible by our own complicity in the system. Even the purchase of a long-desired electric vehicle does not obviate that fact, but indeed underscores it, notwithstanding the considerable courage it took to finally take direct consumer action and buy the car. In the courage of our actions, we mark ourselves in the social and economic order as a person with a hybrid automobile as opposed to a Sport Utility Vehicle (SUV) by driving a car limited to pavement as opposed to an off-road vehicle that has the potential to tear up delicate terrain, and as an owner who, for all the neighbors to see, demonstrates an economic sensibility that coincides with a conservation ethic, as opposed to a sensibility of endless overconsumption… well, one gets the picture.

I can't help but recall a certain vice president who insisted, derisively, that energy conservation may be a fine individual hobby, and a "sign of personal virtue," but would never form the basis for any rational energy policy.[12] No Prius will find him and his kind. Instead, what matters is an instrumental rationality embodied by the state, enabled by state bureaucracies, and defined by unfettered capitalism. That is where real environmental values are presumed to reside. For a set of policy makers, and an even larger set of captains of industry, it is resource use, not resource conservation, which remains the paramount environmental value. In this regard, the ghost of Marx is forever at hand. Means of production and relations of production still matter, as does the social category unpopular in American consciousness, *class*.

Personal virtues are not the appropriate level for rational energy development, apparently. The state, or society, or the market—it is at the level of the collectivity that governments, like forms of capitalism, function. Yet there remains in the United States one abiding symbolic curiosity. In a country that valorizes the individual, the collectivity still rules. Dismissing individual effort, former vice president Dick Cheney thus invoked a telling symbolic armamentarium, implicitly opposing the individual against society—an old, old story. "Austerity" (the realm of the individual), he insisted, was not the issue. "Efficiency" (the realm of the collectivity) provided the path to sustainable energy use. Cheney's efficiency turned into an oil-soaked Gulf of Mexico, requiring—what else?—a massive, social response. One wonders if a bit of austerity would have prevented at least some of the environmental and economic consequences. In any case, the deep ecologists marshal the same general symbolic structure but emphasize the importance of the individual, whose dancing and empirical involvement is where it's at, as opposed to society, where it's not.

Our trusted intuitions, which flow from our deepest yearnings, are subject to a cultural system and to its scaffolding of prior meanings. They do not emerge *ex nihilo* from the dance, no matter how much the barefoot phenomenologists and the deep ecologists wish it were so. Indeed, *not* trusting our intuitions, but setting them side-by-side with alternatives, shows how culturally conditioned they in fact are. Thus, if we fail to understand the differences between traditional ecological knowledge and western science, the Clayoquot Scientific Panel is quick to provide us with proper definitions. Traditional ecological knowledge is all of the following:

> *Holistic*: all things are interconnected and nothing is comprehended in isolation; *Intuitive*: based on deeply held holistic understanding and knowledge; *Qualitative*: knowledge is gained through intimate contact with the local environment, while noting patterns or trends in its flora, fauna, and natural phenomena…*Transmitted intergenerationally by oral tradition*: teaching is accomplished through stories and participation of children in culturally important activities; *Governed by a Supreme Being*: the Creator defines a moral universe with appropriate laws; *Moral*: there are right ways and wrong ways to relate to the environment; *Spiritual*: rooted in a social context that sees the world in terms of social and spiritual relations among all life forms. All parts of the natural world are infused with spirit. Mind, matter, and spirit are perceived as inseparable. Traditional ecological knowledge, in practice, exhibits humility and a refined sense of responsibility; it does not aim to control nature; *Based on mutual well-being, reciprocity, and cooperation*: these promote balance and harmony between the well-being of the individual and the well-being of the social group; *Non-linear*: view time and process as cyclical; *Often contextualized within the spiritual*: may be based on cumulative collective practical and spiritual experience. Traditional ecological knowledge may be revised daily and seasonally through the annual cycle of activities (as required); *Communal*: general knowledge and meaning are shared among individuals horizontally, not hierarchically; and *Promoting of stewardship*: takes a proactive approach to environmental protection and an ecosystem approach to resource management.[13]

The problem is, not all scientists see things with such cooperative good will. For example, from the depths of the 1,200 pages of the 2003 edition of *Wild Mammals of North America*, a caribou biologist plays one set of intuitions off of another: "We must be cautious about TEK perspectives and subject them to rigorous evaluation, much in the manner of Western science (excluding statistical testing)." Why the caution and rigorous evaluation? Well, because First Nations, we are told, don't understand caribou, despite thousands of years interacting with it:

> In Canada, education of Natives to the realities of current caribou use has been and will continue to be a long-term process. It is unlikely, if the educational process is successful, that it will be accomplished without some caribou populations first being extirpated or nearly so. Even a catastrophic loss of caribou would not necessarily convince the Natives of the need to regulate harvests. Because of their beliefs, they most likely would not recognize or accept that they were responsible for the loss.[14]

For their parts, First Nations have often been distrustful of biologists, whose mucking around caribou populations has been interpreted as the cause of population declines. Fastening radio collars to the necks of caribou, chasing them by helicopter, capturing them with nets, sedating them with drugs—all are forms of disrespect to which the caribou respond by going elsewhere.

Biologists don't understand that caribou are non-human persons, with wills and motivations and a deep understanding of their surroundings. They return in their yearly migrations, and allow themselves to be killed, only if they are treated properly. Killing and eating animals is one way to ensure their return. Why? Because they are grateful prey, as told in anthropologist Robert Brightman's intriguing book on human-animal relations among the Cree.

Grateful? How can a biologist trained in objective thinking understand that, let alone incorporate such an idea into wildlife management practices? This is where lists of TEK characteristics fall flat. In order to minimally understand Cree conceptions, our biologist must understand their dreaming and dream images (*pawākan*) through which Crees say they receive knowledge useful in hunting. Such dream images may be of any number of objects or animals—ice, rocks, wolverine, mice, birds—but they are crucial for a hunter's success. Additionally, our biologist must understand cleanliness and pollution, sometimes communicated by the *pawākan* dictating certain practices to follow and avoid.

Proper cleanliness or improper dirtiness influences the success or failure of a hunt, which is to say whether or not an animal will present itself to be killed. The kinds of *pawākan* one dreams also has an effect. The most powerful *pawākan* include bears and various astronomical and meteorological phenomena, the sun and moon, certain stars, clouds, and wind. Dream images presage and guide a successful hunt, which is further influenced by the conduct of the hunter and the will of the prey. Other cultural elements also come into play, for example the prevalent idea that humans and animals are involved in a recurring cycle of gift exchange. Brightman explains:

> Since the soul survives the killing to be reborn or regenerated, the animal does not fear or resent the death. The animals' motivations for participating in these events of killing are figured both in the idioms of love and of interest. Animals may "pity" the hunters who have need of their flesh, and especially is their benevolence evoked when the hunter complies with the conventional objectifications of "respect," treating carcass, meat, and bones in the correct fashion. Conversely, ritual omission or blasphemy angers the animals, who then withhold themselves.

Brightman emphasizes that animal motivations and human motivations are thus completely intertwined, as are the possibilities of any ongoing relationships.

> But the role of the hunter-eater is not that of passive recipient only, and the animals themselves stand to gain from the exchange. Having received the gift

of the animal's body, the hunter reciprocates. Animal souls are conceived to participate as honored guests at feasts where food, speeches, music, tobacco, and manufactured goods are generously given over to them. Hunter and prey are thus successively subject and object in an endless cycle of reciprocities. Ultimately, the roles of human and animal are complementary, for each gives life to the other. The treatment of the remains not only objectifies respect but is said to restore the animal to a living condition.[15]

TEK-savvy biologists, however, once they've come this far, also need to understand how Cree interpretations of the world they inhabit have been altered by historical events, as Crees themselves adjusted to the encroachment of global economies, state bureaucracies, and western science. Historian Richard White has noted that to do otherwise, and to imagine indigenous culture as somehow ahistorical, "can be viewed as an act of immense condescension." Such a viewpoint is not merely condescending but an act of profound empirical wrongheadedness.

As Crees participated in the mid-nineteenth century overkill of beaver and other animals, and witnessed the resulting depopulation, they rethought their own conceptions. No longer could Crees kill as many animals as possible—using only selected parts such as choice portions of meat to eat and hides to sell—and still anticipate a full population return in the following year. By virtue of the simple empirical reality of fewer animals, cause and effect needed to be revised.

Yet wildlife "management" in the form of delayed hunts was itself counterproductive. Any delay meant immediate deprivation. Going hungry was not an option. A reevaluation was needed, which both preserved cultural meanings and altered them in the new context. Crees responded by moving to another area, in the hopes of finding larger populations of game, or by staying put and adopting selective trapping practices and limiting their hunting areas.

Crees faced two threats, one empirical, with a continuing loss of animals, and one symbolic, as prior notions failed to account for the continuing loss. Crees then adapted their symbolic scheme to the new circumstances, as depopulated animals placed their traditional meanings at risk. Animals, as Crees now saw it, demanded that humans hunt selectively. To do otherwise would invite spiritual vengeance from the animals themselves.

But humans needed to do more than selectively hunt their grateful prey. They needed to use the parts of all killed animals in *new* ways, that is, comprehensively. Animals could no longer be "wasted" in anticipation of natural regeneration. All parts of all animals required proper use—for food, clothing, and tool manufacture. Crees now saw themselves, moreover, as capable of influencing animal reproductive potential, whereas before this period of time animals controlled their own numbers.

Selective hunting, comprehensive use of animal parts, and a recognition that humans influenced animal populations were reevaluations of traditional ecological knowledge. Such a shift in understanding "must have constituted a minor scandal in a theology whose dominant postulate was human dependence on animal benefactors," as Brightman puts it. But scandal or not, these cultural changes placed animal depopulations within a symbolic scheme in which animals retained power, spiritual significance, and self-motivation—attributes which remain outside of Western scientific assessment.[16]

Of course, both TEK-savvy biologists and biologically-savvy Crees emerged from a shared history, but not simply as layered cultural artifacts. Symbolism does not result in some sort of sedimentary process of accumulation: dig down far enough through the layers and you'll find strata of traditional ecological knowledge or received biological wisdom. Rather than caking like so much ancient mud, TEK is filtered through Western cultural categories just as Western biology is filtered through Cree categories. It hardly bears saying that the process happens continuously, as cultures interact. The result is not cultural randomness, as if symbolic systems were subject to dynamics of Brownian motion. Nor is it symbolic pastiche, as postmodernists used to say. Crees remain Cree, even as certain members become university-trained biologists, just as Western scientists remain Western scientists, even as certain of them see glimpses of another way of being human.

TEK has spawned, like migrating salmon, a substantial literature in both the biological and social sciences. *Conservation Biology* publishes almost as many articles on TEK as does *Human Organization*. Unlike salmon, however, migrating TEK has strayed far from its natal home, and can be found in diverse management practices, from China to New Zealand to Nunavik, Canada. TEK has become institutionally secure, but it is increasingly apparent that the local cultural knowledge that it purports to convey is hard to discern—at least it remains substantially hidden from managers of natural resources and the bureaucracies in which they work.

Anthropologists have attempted to explain non-Western ecological concepts and practices to Western scientists and bureaucrats, with mixed results. The results are mixed for any number of reasons, including the diverse theoretical and methodological approaches anthropologists employ, bureaucracies variously open to incorporating traditional ecological knowledge into resource management practices, scientists variously willing to consider alternative forms of knowing and engaging the environment, and the simple fact that traditional knowledge itself is dynamic and changing.

The dynamism of indigenous environmental knowledge and use, especially as local peoples adapt to and in many ways adopt Western notions and technologies, remains understudied. As Roy Ellen remarks about such studies generally, what we are left with is "a glut of codification of local environmental knowledge, which curiously undermined our appreciation of the site-specificity, spontaneity and flexibility that are among its fundamental features."[17] What we are left with, in other words, are static lists.

We need to ask a version Bruno Latour's question, *Have we ever been modern?* with the related question, *Have we ever been traditional?* Latour notes that "the idea of a stable tradition is an illusion that anthropologists have long since set to rights." Historians, too, have set to rights the illusion of stability: "The immutable traditions have all budged—the day before yesterday." Latour goes on to say that "one is not born traditional; one chooses to become traditional by constant innovation."[18]

Traditional ecological knowledge in natural resource governance is both chosen and imposed. TEK is a mixed form of cultural innovation that ignores history, generating the appearance of stability. Such stability is then put to some purpose, or rather to multiple purposes, since the notion of tradition is worked into the diverse practices of many different participants, from politicians to wildlife managers to anthropologists to native peoples to law enforcement personnel…to scientists. The constant invention of tradition is articulated within the minutia of bureaucratic discourse and played out in the vastness of managed landscapes, instantiated in the daily practices of contemporary hunter-gatherers and demonstrated in the historical dioramas of National Park Service interpretive displays. Thus do we see traditions appearing to outlast the cultural bases of their *raison d'être*, motivated by other social forces that seek to stabilize them.

Notes

1 Mary Douglas, "Traditional Culture—Let's Hear No More about It," in Vijayendra Rao and Michael Walton, eds., *Culture and Public Action* (Stanford, CA: Stanford University Press, 2004), 88.

2 World Commission on Environment and Development (Brundtland Commission). *Our common future* (Oxford: Oxford University Press, 1987), 115.

3 "Sustainable Ecosystem Management in Clayoquot Sound: Planning and Practices," The Clayoquot Sound Scientific Panel (April 1995), 1.

4 "First Nations' Perspectives Relating to Forest Practices Standards in Clayoquot Sound," The Clayoquot Sound Scientific Panel (March 1995), 14.

5 James C. Scott, *Seeing Like a State: How Certain Schemes to Improve the Human Condition Have Failed* (New Haven, CT: Yale University Press, 1998), 331.

6 "First Nations' Perspectives Relating to Forest Practices Standards in Clayoquot Sound," 5.

7 Roy A. Rappaport, *Ritual and Religion in the Making of Humanity* (Cambridge: Cambridge University Press, 1999), 459.

8 Global Assessment Report on Biodiversity and Ecosystem Services https://ipbes.net/global-assessment. Annie Sneed, "What Conservation Efforts Can Learn From Indigenous Communities," *Scientific American* (May 29, 2019), provides an accessible summary.

9 Bill Deval and George Sessions, *Deep Ecology* (Salt Lake City, UT: Peregrine Smith Books, 1985), 7.

10 Erik Gómez-Baggethun, Rudolf de Groot, Pedro L. Lomas, and Carlos Montes, "The History of Ecosystem Services in Economic Theory and Practice: From Early Notions to Markets and Payment Schemes," *Ecological Economics* 69 (April 2010).

11 www.fs.fed.us/ecosystemservices/About_ES/index.shtml

12 James Carney and John F. Dickerson, "The Rocky Rollout of Cheney's Energy Plan," *Time* (May 19, 2001).

13 "First Nations' Perspectives Relating to Forest Practices Standards in Clayoquot Sound," 14–15.

14 Frank L. Miller, "Caribou (Rangifer tarandus)," in George A. Feldhamer, Bruce C. Thompson, and Joseph A. Chapman, eds., *Wild Mammals of North America: Biology, Management, Conservation*, second ed. (Baltimore, MD: Johns Hopkins University Press, 2003).

15 Robert Brightman, *Grateful Prey: Rock Cree Human-Animal Relationships* (Regina: Canadian Plains Research Center, 2002), 187–188.

16 Brightman, *Grateful Prey*, 308.

17 Roy Ellen, Forward, *Landscape, Process and Power: Re-evaluating Traditional Environmental Knowledge*, Serena Heckler, ed. (New York, NY: Berghahn Books, 2009), xiv.

18 Bruno Latour, *We Have Never Been Modern*, Catherine Porter, trans. (Cambridge: Harvard University Press, 1993), 75, 76.

11

THE DHARMA OF NATURE

In the last several decades, notions of ecological change have become central components of a new environmental awareness. Prior ideas of ecological balance and ordered ecological succession have given way to ideas of contingency, of unexpected and unpredictable ecological modification. Risk and uncertainty, rather than predictability and control, should be at the core of land management. But as Max Weber first clearly saw, and as this book shows, bureaucracies with their ordered rankings of certified experts subject to managerial oversight often insist on predictability and control in the face of a contradictory world. Many of the earth's recent and contingent environmental changes have been the result of human activity: global warming, deforestation, and the loss of biodiversity, to cite common examples, are directly attributable to human influence. Human-caused environmental changes, however, are not new. Humans, through hunting and fishing practices, farming techniques, widespread use of grassland and shrub-land fires, and their own social interactions, have substantially altered ecosystems across the planet for many thousands of years, sometimes dramatically.[1]

"Human beliefs about the nature of ecology," writes Lawrence E. Sullivan, "are the distinctive contribution of our species to the ecology itself."[2] The challenge facing all of us is to recognize the pervasive human influence in the natural world, and to act, or to choose not to act, based on that recognition. As I noted at the outset, Western-style land management, governed by economic rationality and state bureaucracies, may not result in improved environments. If instead we consider the widest possible human influence—all those human beliefs about the nature of ecology—and not just our own cultural, economic, and bureaucratic forms, then how do we produce ideal environments for future generations? The related question—who gets to choose?—presumably lies in the realm of politics.

Framed in this way, environmental values take on added importance. Values inform perceptions of environmental change, the need to address such changes,

DOI: 10.4324/9781003297444-14

and the manner in which environmental improvements should be effected. They also provide justifications for actions taken or proposed, as well as justifications for inaction. But the identification and analysis of environmental values is particularly problematic. Such values are deeply enmeshed in many realms of human life—practical, economic, religious, material, governmental, and recreational. Like those realms, environmental values have been subject to substantial and ongoing change. They are not static, no more than environments are static. Nor do environmental values always find articulation in discrete, easily recognizable domains of meaning. They are parts of many such domains, at different levels of explicitness, and the level of explicitness is itself subject to historical change.

In addition, environmental values are increasingly the site for public contestation, which creates its own dynamic of change. Notions of morality, proper governance, democracy, economic well-being, sustainable development, local/global relations, wilderness preservation, multiple use, animal rights, food security, and so on, are part of the many discourses of environmental values expressed in diverse public forums—from local potluck dinners hosted by members of Earth First!, to mass street-level protests of the World Trade Organization, to global conferences on the fate of our planet. Land management agencies, however, try to domesticate the diversity of environmental values, often turning them into "ecosystem services." But such domestication is just as actively resisted by those with contrary ideas.[3]

Tug on the thread of any environmental value—sustainable fisheries, say, or healthy forests, or potable water, or wildlife preservation, or an experience of the sublime—and what unravels is a skein of politics, bureaucracy, technology, culture, science, economics, nature, and individual agency.[4] The results of the tug suggest that these domains are not as securely separated as their labels imply. And this is precisely the difficulty for an analysis of environmental values: to discover where they are located, how they are expressed, and why once they are expressed they may become part of the expected and natural order of things, which is to say they are forgotten as human creations. Over-thick forests, dead pools, declining wildlife, exhausted fisheries: all of these are human created, the result of decisions of those who came before us.

It is often hard to recognize our contemporary world as the product of past decisions; instead, we take our world as a given. Yet especially in the context of old decisions and enemy ancestors, environmental values lead labyrinthine lives, winding through many domains of meaning and pragmatic experience, some with historical sources, others with sources closer at hand. Tug the thread hard enough and all areas of human life will be found connected.

Anthropologist Roy Rappaport, who was instrumental in the development of ecological anthropology in the 1960s and 1970s, thought that Western values and economic systems are badly maladaptive. Western economies contain the engines of their own destruction, he argued, while Western values allow and encourage environmental degradation for economic gain. Rappaport believed that there were no viable social mechanisms to rein in the maladaptive behavior, to forestall or stop the use of nonrenewable resources upon which Western economic systems depend.[5]

In the last several decades, however, there have emerged a number of increasingly vocal and powerful social groups that attempt to change the very policies and practices that allow environmentally destructive behavior to continue. Greenpeace is a well-known international example. Examples of smaller coalitions of activists seeking change include the Friends of Nature in China, Acción Ecológica in Ecuador, the Ashoka Trust for Research in Ecology and the Environment in India, the Friends of the Earth in Japan, the Southern Utah Wilderness Alliance in the United States, and the Forest Action Network in Canada. There are now many hundreds, perhaps many thousands, of such groups scattered across the planet.

These emerging social groups, some of which have become well-established, represent a diverse array of environmental values. They are comprised of protest groups and political parties, neighborhood organizations, and virtual communities held together by the Internet. Their concerns may be local, focused on salmon restoration in a single river system, or global, focused on the causes and consequences of worldwide climate changes. Their environmental values take a number of forms, some of which articulate an ecocentric vision of environmental stewardship that insists upon recognizing the fundamental integrity of ecosystems quite apart from any human concerns—the so-called deep ecology movement. Others promote individual experience as the key element in the formation of environmental values, in which spiritual connections to the natural world, or philosophies emphasizing direct perceptual experience, are of paramount importance. Still others see links between social organization, democratic processes, and environmentalism as comprising a foundational set of concerns, out of which environmental values flow.[6]

In addition to environmental activists, local community members are recognizing their roles in promoting economic and other behaviors that are less environmentally destructive. Their environmental values appear to be no less diverse than those of outspoken environmentalists, and no less difficult to describe and analyze as emergent phenomena. In some ways, community involvement in local environmental issues may have greater and longer lasting effects than the involvement of activists, as community members themselves incorporate environmentally sensitive practices into their daily lives and into their political discussions. But the relationship between activist and community is nonetheless important. As local concerns are dealt with locally, they are often brought to the attention of the wider public through the involvement of environmental groups. The converse is also true: environmental groups provide education and material resources to aid local communities with their environmental problems. Local, national, and international groups are becoming increasingly intertwined as environmental values achieve greater popular currency.

Yet in the historical process that brings disparate groups together, the currency itself—environmental values—does not maintain the same meanings across the planet. As globalization continues, it is matched by local differentiation, in values as in other elements of culture. Anthropologist Marshall Sahlins, surveying the planet-wide cultural scene, has this to say:

Consider again this surprising paradox of our time: that localization develops apace with globalization, differentiation with integration; that just when the forms of life around the world are becoming homogeneous, the peoples are asserting their cultural distinctiveness….The short answer to the paradox is… 'resistance.' Problem is, the people are not usually resisting the technologies and 'conveniences' of modernization, nor are they particularly shy of the capitalist relations needed to acquire them. Rather, what they are after is the indigenization of modernity, their own cultural space in the global scheme of things….Hence what needs to be recognized is that *similitude is a necessary condition of the differentiation.*[7]

Sahlins's last point may be a bit obscure: the more similar we become the more different we wish to be. But consider one small example of local people making their own cultural space in the global scheme: at the airport in Port Hardy, on the northern tip of Vancouver Island, Canada, a large banner was, and may still be, displayed across an airplane hangar for all airline passengers to see; it reads: THIS IS A GREENPEACE FREE ZONE. As I was told in 1998, Port Hardy had enacted a local ordinance prohibiting the sale of gasoline to Greenpeace planes and boats, in protest of Greenpeace's involvement in slowing timber harvesting upon which local economies depend. One person's environmental value is another person's lost job. As a bumper sticker in the State of Oregon proclaimed: "Are You an Environmentalist, or Do You Work for a Living?"[8]

Or consider a more complex example from the highland Uva region of Sri Lanka. Nireka Weeratunge compared the English and Sinhalese versions of the 1992 Sri Lankan *Citizens' Report on Environment and Development*. She notes that the Sinhalese phrases *sobadahamata anukula*, which she glosses as "compatible with the *dharma* of nature," and *sobadahama ha yavunu*, "connected to the *dharma* of nature," are both rendered in English as "in harmony with nature." Weeratunge points out, however, that "neither 'harmony' nor 'nature' is commensurable with local terms," and that in the translation to English, meanings are both added and lost.

Harmony, nature, and *dharma* are all complex terms, the adequate translations of which require careful glosses and contextual examples. An awareness of the nuances of local meaning, and an understanding of the inevitable shifts in meaning when words expressing environmental values are translated across languages, are crucial for an analysis of environmental values. Weeratunge argues that the idea of harmony runs counter to the local notion of *kaliyugaya*, which posits a succession of order and disorder directly associated with human morality:

The kaliyugaya discourse sees human beings as both causing and inevitably subject to disorder because they were born, according to their *dharma*, during a chaotic period of time. Thus, present human beings are paying for their own lapses and those of their ancestors. While a fundamental premise of environmentalism is that people who are aware of the crisis and care about the environment can do something to stop environmental destruction, the notion of *kaliyugaya* provides a doomsday prophecy that the current generation can do precious little. In other words, the efforts of activists are in vain, given the immensity of the forces of decline and destruction and the fact that they have been born during the wrong time in history.[9]

Of course, an awareness and understanding of local cultural nuances, or even of fundamental cultural premises, may not be needed for environmental discourse itself: Sri Lankans after all appear to have no qualms about the translations they make from Sinhalese into English. The result is not necessarily the obliteration of local meaning—*dharma* somehow becomes harmony—but rather the reworking of such meanings in the new context.

A recognition of cultural specificity, of the sort many anthropologists are rightly concerned with, may not obtain as the meanings from different cultural traditions are transposed in environmental discourse, even as Western environmentalists continue to mistakenly believe that life in harmony with nature is occurring among Sri Lankans.

Nor should this come as much of a surprise. The adoption of meanings from other cultural traditions is one of the characteristics of Western environmental discourse; doubtless the reverse occurs as well. Jainism, Buddhism, and various American Indian beliefs, among many others, have all been subject to incorporation into the discourse of contemporary Western environmentalism, as that environmentalism attempted to distance itself from the cultural status quo.[10]

Such an incorporation of other meanings, and such a construction of symbolic difference, rarely preserves linguistic nuance, and almost never explicates in any detail the cultural and historical contexts from which alternative environmental meanings initially derived. What frequently happens in such discourse is the distillation of meaning into a form that critiques dominant social/environmental relations. In the political context of environmental movements, new meanings emerge, although they may only superficially resemble the source meanings. A set of cultural and political problems often arise when such meanings, grounded in Western notions (even if originating elsewhere), are accorded greater standing than the alternative meanings themselves, as if indigenous values are provisional and easily subject to modification to bring them in line with Western values. One example is the fluctuating critique of "conservation" and "nature."

"We are not conservationists," Nicanor Gonzalez, a Kuna Indian, asserts about indigenous peoples, in dispute of the popular misapprehension of American Indian belief. "We aren't nature lovers." Gonzalez goes on to say, "What I have understood in talking with the indigenous authorities, indigenous groups, and individuals is that they are familiar with the laws of nature. They aren't conservationists; rather, they know how to interrelate humans and nature. This is the basic principle of indigenous people: interrelating the exchange and communication between Mother Earth and indigenous communities. Therefore, when indigenous people kill or hunt animals, or extract a medicinal plant, it is sacred. It is as if they are asking permission from the earth, from Mother Nature. This is not conservation—it is a contact with nature, communication with nature, communication between nature and person." Buzz Cobell, a member of the Blackfeet tribe, has a different perspective: "We've always managed—we just didn't call it that. The difference is, now we've been acculturated to the point where we approach wildlife management like the federal and state governments do."[11]

Questions of translation are clearly central to the study of global environmental discourse. It is important not to assume similarity of meaning even if the same terms are on every environmentalist's lips. The problem is that meaning is always subject to a cultural system already in place.

Meanings are inevitably mediated by prior structures of significance. In this there should be little cause for wonder. For how else could there be any meaning? Meanings do not move freely, although they may in the age of the Internet move quickly.

At the same time, if our focus is on bureaucratic structure and function, and not on a diversity of cultural values, then Sahlins's formula needs to be reversed: differentiation requires similitude. This is what Weber saw as central to bureaucracy: a rigid division of labor, a quest for optimal efficiency, the predictability of rules, a pathological secrecy, and the elimination of the human. In such systems, human difference is poorly tolerated. Similarity is preferred if not enforced. And yet, as I've tried to show, bureaucracies are considerably more mixed than Weber foresaw, and Sahlins's point about cultural difference is one I've highlighted throughout. Constraining difference, land management bureaucracies confront difference, and indeed benefit from it.

The Wild Garden

Part of the analytical difficulty encountered in *Nature and Bureaucracy* is that science, policy, bureaucratic knowledge, and local or traditional knowledge are different modes of knowing and engaging the world. They do not always or even frequently coincide with or directly inform one another. They may well be, in some circumstances, antithetical modes of knowing. But they are also implicated in one other, often in unexpected ways. Anthropologist Bruno Latour, describing a newspaper report on ozone depletion, notes:

> The same article mixes together chemical reactions and political reactions. A single thread links the most esoteric sciences and the most sordid politics, the most distant sky and some factory in the Lyon suburbs, dangers on a global scale and the impending local elections or the next board meeting. The horizons, the stakes, the time frames, the actors—none of these is commensurable, yet there they are, caught up in the same story.[12]

Latour describes what may be thought of as the opposite of Weeratunge's problematic. He is interested in how incommensurables become linked, Weeratunge with how they are transposed. Both linking and transposition result in complex reworkings of meaning. The difficult question to answer in both circumstances is how, and by what social pathways, environmental values in science and policy making are reflected at local levels or, conversely, how local considerations of environments can be reflected in the environmental values of scientists and policymakers and other land management bureaucrats. The former question has been the concern of scientists attempting to find ways to better inform the public and policymakers about environmental problems, such as climate change. The latter question has been the concern of those trying to incorporate non-Western knowledge into Western scientific understanding and policy decisions—the dharma of nature. These attempts also result in complex reworkings of meaning, as described in various chapters of this book.

How can such meanings be understood, mixing and linking as they do different cultures, histories, natures, political contexts, scientific assessments, folk knowledge, bureaucratic hierarchies, and personal experience? Intellectual practices that focus on power, or discourse, or nature, or bureaucracy are all insufficient to adequately characterize these meanings. The focus on power relations ignores the natural world. The focus on discourse pretends that all meaning is a product of or is conducive to language. A scientific focus on nature, purified of the social, results in an attention of the natural world that masks cultural meanings. An institutional focus on bureaucracies tends to eliminate the human. These foci in fact reinforce a Western ontology which posits a split between humans and the natural world.

One way out of the morass of contemporary intellectual fashion is to follow the thread that connects politics, language, culture, bureaucracy, and nature and to conceive the object of study as "nature-culture," as Latour has it, and to investigate nature-culture in the West in the same anthropological fashion as elsewhere, which is to say holistically—an admittedly unfashionable approach in the fragmented world we inhabit. Perhaps environmental values are prime candidates for such investigations because they cross and link so many different and varied domains, from cosmology to science to democracy to nature to personal agency to technology to culture to bureaucracy, and back again. The translations and reworkings of meanings that occur across linguistic and cultural boundaries, the networks that form and dissolve between different social groups, and the scientific determinations of the natural world that inform both—all are central to our modern world, all are mediated, in some fashion, by bureaucratic structures. And none of them stand securely alone.

We end where we began, with a wild garden. We may be in the age of total bureaucratization, but we control our cultural meanings, our messages, our gardens. "Thwart institutional cowardice," as filmmaker Werner Hertzog forcefully and playfully insists in *A Guide for the Perplexed*.[13] I've tried to show how, by imagining an ordered natural world, and by imposing rules to enforce that imagined order, land management bureaucracies are themselves vast culturally constructed forms. These forms, and the cultural meanings associated with them, are diverse. In Alaska, native peoples, anthropologists, and fish were caught up in various bureaucratic structures, as they all pursued their own purposes. In North American forests, bureaucracies struggled to alter natural fire regimes for commercial profit—and to deeply contra-dictory ends. Along the east coast of North America, wild Atlantic salmon at the edge of extinction showed the failure of environmental policies and highlighted the limited ability of national and state bureaucracies to bring back the near-dead. In the desert southwest, pumping water uphill to slake the thirst of desert cities came at a cost, the eventual inability of those cities to maintain their growth.

Neither entirely managed, nor entirely wild, our public lands and our public resources are increasingly cultured by a dominant humanity. As we continue to sort out the consequences of the decisions of our predecessors, and as we con-tinue to make our own decisions about our shared planetary resources, my hope is to avoid becoming an enemy ancestor to our descendants. We have plenty of our own enemy ancestors whose decisions were less future-oriented than they should have been. Their decisions have often made our lives less secure than they could be. I wish to change both. Nudging, sometimes pushing, the bureaucracies within which I work is a start.

Notes

1 For a review of concepts of ecological change see Daniel B. Botkin, *Discordant Harmonies: A New Ecology for the Twenty-First Century* (New York, NY: Oxford University Press, 1990). On the dynamic nature of ecosystems, see Claudia Pahl-Wostl, *The Dynamic Nature of Ecosystems: Chaos and Order Entwined* (New York, NY: John Wiley, 1995). For studies of human impacts on local ecosystems see Carole L. Crumley, ed., *Historical Ecology: Cultural Knowledge and Changing Landscapes* (Santa Fe, NM: School of American Research Press, 1994), Jeanne X. Kasperson, Roger E. Kasperson, and B.L. Turner II, eds. *Regions at Risk: Comparisons of Threatened Environments* (New York, NY: United Nations University Press, 1995), Shepard Krech III, *The Ecological Indian: Myth and History* (New York, NY: Norton & Company, 1999), Mark J. McDonnell and Stewart T.A. Pickett, eds., *Humans as Components of Ecosystems: The Ecology of Subtle Human Effects and Populated Areas* (New York, NY: Springer-Verlag, 1993), J.R. McNeill, *Something New Under the Sun: An Environmental History of the Twentieth-Century World* (New York, NY: W.W. Norton & Company, 2000), Charles L. Redman, *Human Impact on Ancient Environments* (Tucson, AZ: University of Arizona Press, 1999), Joachim Radkau, *Nature and Power: A Global History of the Environment* (Cambridge: Cambridge University Press, 2008), Andrea Wulf, *The Invention of Nature: Alexander Humboldt's New World* (New York, NY: Vintage Books, 2016). This literature is large and growing.
2 Lawrence E. Sullivan, Preface, in John A. Grim, ed., *Indigenous Traditions and Ecology: The Interbeing of Cosmology and Community* (Cambridge: Harvard University Press, 2001), xi.
3 See Sarah F. Trainor, "Realms of Value: Conflicting Natural Resource Values and Incommensurability," *Environmental Values* 15 (2006).
4 Sustainable development is thought by some to be the obvious and natural solution to environmental degradation. But sustainable development is itself a human creation riddled with contradictions. See Shiv Visvanathan, "Mrs. Bruntland's Disenchanted Cosmos," *Alternatives* 16 (1991); see also Arturo Escobar, *Encountering Development: The Making and the Unmaking of the Third World* (Princeton, NJ: Princeton University Press, 1995). With a focus on energy flows and material transfers, "Industrial Ecology," despite its systems perspective and its focus on sustainability, typically fails to adequately consider

cultural elements of the system—values, laws, cultural conceptions of nature, etc. See Jouni Korhonen, "Industrial Ecology for Sustainable Development: Six Controversies for Theory Building," *Environmental Values* 14 (2005).

5 Roy A. Rappaport, *Ecology, Meaning, and Religion* (Richmond, CA: North Atlantic Press, 1979), and *Pigs for the Ancestors: Ritual in the Ecology of a New Guinea People*, second ed. (Long Grove, IL: Waveland Press, 1984). The central analytical failing of Rappaport's work is his adoption of notions of ecological homeostasis or dynamic equilibrium, some of which he appears to downplay in his last book, *Ritual and Religion in the Making of Humanity* (Cambridge: Cambridge University Press, 1999). In this work, the focus is less on ecosystems and more on culture.

6 On deep ecology, see Arne Naess, "The Shallow and the Deep, Long-Range Ecology Movement: A Summary," *Inquiry* 16 (1973), Bill Devall and George Sessions, *Deep Ecology: Living as if Nature Mattered* (Salt Lake City, UT: Peregrine Smith, 1985), George Sessions, ed., *Deep Ecology for the Twenty-First Century: Readings on the Philosophy and Practice of the New Environmentalism* (Boston, MA: Shambhala Publications, 1995). Stephen Avery provides a critique of the non-anthropocentric ethic associated with deep ecology, "The Misbegotten Child of Deep Ecology," *Environmental Values* 13 (2004). For discussions of phenomenology and environmental values, see David Abram, *The Spell of the Sensuous: Perception and Language in a More-Than-Human World* (New York, NY: Pantheon Books, 1996), and Linda Nash, "The Changing Experience of Nature: Historical Encounters with a Northwest River," *Journal of American History* (2000). For a discussion of democracy and environmentalism, see Michael Mason, *Environmental Democracy* (London: Earthscan Publications, 1999).

7 For an intriguing essay that touches on current local cultural differentiation in the face of globalization, see Marshall Sahlins, "Two or Three Things That I Know About Culture," *Journal of the Royal Anthropological Institute* (N.S.) 5 (1999). See also various essays in Sahlins, *Culture in Practice: Selected Essays* (New York, NY: Zone Books, 2000).

8 For a discussion of work and environmentalism, see Richard White, "'Are You an Environmentalist, or Do You Work for a Living?': Work and Nature," in *Uncommon Ground: Rethinking the Human Place in Nature*, William Cronon, ed. (New York, NY: W.W. Norton, 1996).

9 Nireka Weeratunge, "Nature, Harmony, and the *Kaliyugaya*," *Current Anthropology* 41: 249–268 (2000), 252, 259.

10 For examples, see Mary Evelyn Tucker and John A. Grim, eds., *Worldviews and Ecology: Religion, Philosophy, and the Environment* (London: Bucknell University Press, 1994), John A. Grim, ed., *Indigenous Traditions and Ecology: The Interbeing of Cosmology and Community* (Cambridge: Harvard University Press, 2001), Pablo Piacentini, ed., *Story Earth: Native Voices on the Environment* (San Francisco, CA: Mercury House, 1993).

11 Nicanor Gonzalez, "We Are Not Conservationists," *Cultural Survival* 16 (1992), 45. Michelle Niihuis, "Wildlife Management Blossoms on the Reservation," *High Country News* 21 (February 2001), 10.

12 Bruno Latour, *We Have Never Been Modern*, C. Porter, trans. (New York, NY: Harvester Wheatsheaf, 1993).

13 Werner Herzog, *A Guide for the Perplexed: Conversations with Paul Cronin* (New York, NY: Farrar, Straus, and Giroux, 2014).

INDEX